AF898938

Lecture Notes on Multidisciplinary Industrial Engineering

Series Editor

J. Paulo Davim, Department of Mechanical Engineering, University of Aveiro, Aveiro, Portugal

"Lecture Notes on Multidisciplinary Industrial Engineering" publishes special volumes of conferences, workshops and symposia in interdisciplinary topics of interest. Disciplines such as materials science, nanosciences, sustainability science, management sciences, computational sciences, mechanical engineering, industrial engineering, manufacturing, mechatronics, electrical engineering, environmental and civil engineering, chemical engineering, systems engineering and biomedical engineering are covered. Selected and peer-reviewed papers from events in these fields can be considered for publication in this series.

More information about this series at http://www.springer.com/series/15734

Inderdeep Singh · Pramendra Kumar Bajpai ·
Kuldeep Panwar
Editors

Trends in Manufacturing Processes

Select Proceedings of ICFTMM 2018

Springer

Editors
Inderdeep Singh
Indian Institute of Technology Roorkee
Roorkee, Uttarakhand, India

Pramendra Kumar Bajpai
Netaji Subhas University of Technology
New Delhi, Delhi, India

Kuldeep Panwar
Delhi Technical Campus
Greater Noida, Uttar Pradesh, India

ISSN 2522-5022 ISSN 2522-5030 (electronic)
Lecture Notes on Multidisciplinary Industrial Engineering
ISBN 978-981-32-9098-3 ISBN 978-981-32-9099-0 (eBook)
https://doi.org/10.1007/978-981-32-9099-0

© Springer Nature Singapore Pte Ltd. 2020
This work is subject to copyright. All rights are reserved by the Publisher, whether the whole or part of the material is concerned, specifically the rights of translation, reprinting, reuse of illustrations, recitation, broadcasting, reproduction on microfilms or in any other physical way, and transmission or information storage and retrieval, electronic adaptation, computer software, or by similar or dissimilar methodology now known or hereafter developed.
The use of general descriptive names, registered names, trademarks, service marks, etc. in this publication does not imply, even in the absence of a specific statement, that such names are exempt from the relevant protective laws and regulations and therefore free for general use.
The publisher, the authors and the editors are safe to assume that the advice and information in this book are believed to be true and accurate at the date of publication. Neither the publisher nor the authors or the editors give a warranty, expressed or implied, with respect to the material contained herein or for any errors or omissions that may have been made. The publisher remains neutral with regard to jurisdictional claims in published maps and institutional affiliations.

This Springer imprint is published by the registered company Springer Nature Singapore Pte Ltd.
The registered company address is: 152 Beach Road, #21-01/04 Gateway East, Singapore 189721, Singapore

Preface

The Trends in Manufacturing Processes book covers a variety of topics from diverse filed of manufacturing. The contents of the edited book is based on papers presented at the First International Conference on Futuristic Trends in Materials and Manufacturing (ICFTMM-2018) held in Delhi Technical Campus, 28/1, Knowledge Park-III, Greater Noida, Uttar Pradesh 201306, India, on 26–27 October 2018. The topics include research findings in the field of conventional and non-conventional manufacturing processes. Conventional manufacturing, advanced manufacturing processes, and various fabrication processes of advanced materials have been covered in this book. This book focuses on the field of manufacturing engineering, which is a branch of engineering concerned with the perceptive and application of engineering actions in manufacturing processes and production methods. Manufacturing engineering requires the ability to plan the practices of manufacturing; to research and to develop tools, processes, machines, and equipment; and to integrate the facilities and systems for producing a quality product with the optimum expenditure of capital.

This book covers in-depth advanced manufacturing processes, and optimization of various process parameters and performance parameters used in the manufacturing processes. It is expected that this book will be suitable for the students, academicians, engineers, and researchers working in this field.

Roorkee, India
New Delhi, India
Greater Noida, India

Inderdeep Singh
Pramendra Kumar Bajpai
Kuldeep Panwar

Contents

About the Editors

Dr. Inderdeep Singh completed his bachelors in mechanical engineering in 1998 from the National Institute of Technology Hamirpur, followed by his masters and Ph.D. (in composites materials) from IIT Delhi in 2000 and 2004, respectively. His doctoral research work was awarded by the Foundation for Innovation and Technology Transfer as the 'Best Industry Relevant Ph.D. Project' during the session 2004–05. Dr. Singh is currently Associate Professor in the Mechanical and Industrial Engineering Department at IIT Roorkee. He has published 180 articles in international and national journals and conferences. He has authored a book titled "Primary and Secondary Manufacturing of Polymer Matrix Composites" and has also contributed 15 book chapters in the field of composite materials. His current research focuses on developing fully biodegradable green composites.

Dr. Pramendra Kumar Bajpai is an Associate Professor in the Division of Manufacturing Processes and Automation Engineering at Netaji Subhas Institute of Technology, New Delhi. He completed his Ph.D. from the Department of Mechanical and Industrial Engineering, Indian Institute of Technology, Roorkee. His major research areas include processing of polymer composites and green composites, machining aspects of composites, and FEM modeling of composites. He has published several research articles in journals of repute. He has also contributed many book chapters published by noted publishing houses. He is currently working as an editor of two contributed volumes on processing of polymer composites.

Dr. Kuldeep Panwar is currently working as Associate Professor and Head of Mechanical and Automation Engineering Department at Delhi Technical Campus, Greater Noida. He completed his Bachelors of Engineering from B.T.K.I.T, Dwarahat (erstwhile KEC) in 2009, and his Masters in Thermal Engineering in 2012. He completed his Ph.D. in Mechanical Engineering from G. B. Pant University of Agriculture and Technology, Pantnagar in 2016. His doctoral research area was analysis of thermal and resistance characteristics of fixed-bed regenerator with small D/dp ratio. He has published more than 30 research articles in peer-reviewed international journals and conferences. He has one design patent to his credit. He has authored a book titled Effect of Insolation on Efficiency of Solar Plate Collectors and has also contributed as an editor in 4 conference proceedings.

Chapter 1
Optimization-Related Studies on the Operational Parameters of Friction Stir Welding: An Overview

Manish Kumar Singh, Rajesh Kumar Porwal and Sanjay Mishra

Abstract Friction stir welding technique is one of the most emerging, environmental-friendly and rapid-growing solid-state joining technique based on heat generated due to friction and application of pressure. This technique is generally used to join materials that have wide range of industrial applications and that are difficult to weld by fusion welding process, like aluminum, copper, nickel. In the past, researchers have searched a number of approaches to improve joint properties by analyzing process parameters that affect the mechanical and microstructural characteristics of the joint. The experimental and theoretical studies show that the weld joint properties can be improved considerably by proper selection of process parameters. This article shows an overview of various optimization techniques that can be used to find the optimum welding operational parameters during the joining of similar and dissimilar materials.

Keywords Friction stir welding · Operational parameters · Optimization

1.1 Introduction

The applications of aluminum and copper, categorized as soft materials, are mostly in the aircraft, shipping, defense and transportation industries as they possess certain properties, like being lightweight, having a high corrosion resistance, good mechanical properties and a good strength-to-weight ratio. Coalesce of these soft materials is difficult using fusion welding as there is an establishment of brittle inter-metallic compound at the interface of weld joint that results in decrease in the mechanical and microstructural properties. This problem was overcome by using FSW, in which metallic bond is produced at a lower temperature. The weld strength is stronger in

M. K. Singh · R. K. Porwal (✉)
Faculty of Mechanical Engineering, Shri Ramswaroop Memorial University, Lucknow, India
e-mail: porwal.me@srmu.ac.in

S. Mishra
Department of Mechanical Engineering, Madan Mohan Malaviya University of Technology, Gorakhpur, India

© Springer Nature Singapore Pte Ltd. 2020

I. Singh et al. (eds.), *Trends in Manufacturing Processes*,
Lecture Notes on Multidisciplinary Industrial Engineering,
https://doi.org/10.1007/978-981-32-9099-0_1

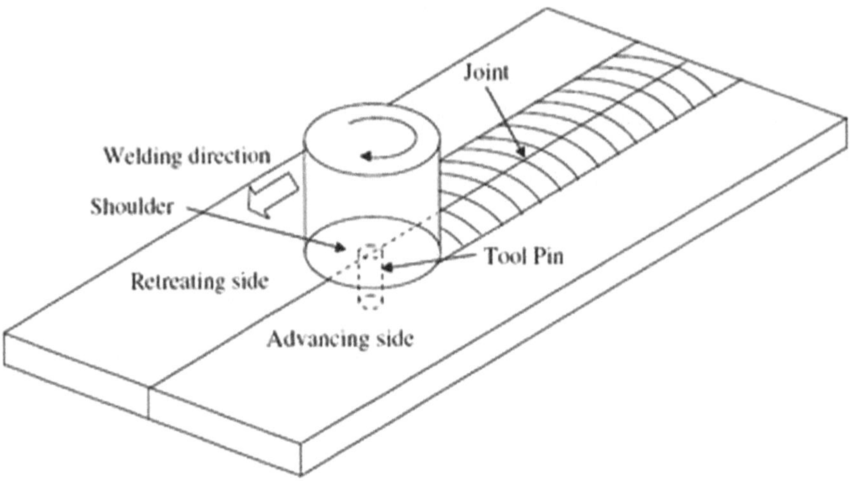

Fig. 1.1 Schematic view of FSW [6]

FSW because there is no melting of base metal; thus the heat-affected zone is narrower in this process compared with the fusion welding. Although FSW was initially developed to join aluminum and its alloy, recently it is applied successfully and effectively to join other metals and materials [1–3]. There are some other materials which were joined successfully by the FSW technique, such as magnesium, nickel, titanium, steels and metal matrix composites. This technique can also be used to join dissimilar and thermally un-weldable materials [4].

Friction stir welding was first introduced by Thomas and Nicholas at The Welding Institute, Cambridge, UK in 1991. It uses a non-consumable rotating tool having a pin below the shoulder. Both the pin and shoulder of tool contributed to heat generation. Pin mixes the base materials while shoulder controls the material flow. The movement between the tool and workpiece produces heat by means of friction which softens the base materials. Soften material flow attributable to major plastic deformation and is united in order to make a successful joint. The schematic view of friction stir welding is shown in Fig. 1.1. There are two welded sides in the FSW; first one is the advancing side in which the direction of rotating tool and the weld are in the same direction, while the second one is known as retreating side, in which the direction of rotating tool and weld direction are in a different direction [5–7].

1.2 Review Related to Experimental and Theoretical Studies of FSW

This literature is a review of the experimental and theoretical studies performed by the researchers across the globe, which shows the effect of various process parameters, such as rotational speed, travel speed and tool geometries, on the weld joint properties. For example, Liu et al. performed FSW on 6061-T6 Al alloy having thickness of 0.63 cm, with RS varying from 300 to 1000 rpm and traverse speed compassed from 0.15 to 0.25 cm/s. The hardened carbon steel as a tool material having 0.63 cm pin diameter along with 1.25 cm shoulder diameter was used for their experimental study. They compared weld joint mechanical properties with the base metal and it was found that the residual hardness varied from 55 VHN at the top of weld to 65 VHN at the bottom of the weld as compared with the workpiece hardness, which varied from 85 to 100 VHN, as shown in Fig. 1.2 [8]. Weld joint strength was investigated by Saeid et al. during the FSW of 1060 Al alloy with pure copper in lap configuration. Quenched and tempered steel as a tool material having a 15 mm shoulder diameter with threaded pin profile was used for their experimental investigation. They concluded that the maximum tensile shear strength of the joint was achieved at 95 mm/min welding speed. They also observed cavity defect at 118 and 190 mm/min welding speed [9]. Yabuuchi et al. performed FSW on oxide dispersion strengthened steel to investigate the effect of RS on mechanical properties and microstructure of the weld joint. They demonstrated that there was a reduction in stirred weld zone hardness at all RS because of recrystallization lured by temperature generation during the joining process. Furthermore, recrystallization was followed by a modification in grain structure from prolonged to isotropic grains and grain size was increased with an increase in RS [10]. Fratini et al. studied the impact of work material characteristics on plasto-mechanics during the FSW of Al alloys 2024-T4 and 6082-T6 in T-joints configuration. The authors used a special design fixture to determine the process conditions effect, metallurgical and mechanical behavior of joints. The thermal and plastic flow fields of workpiece materials were calculated and compared by using numerical methods. Their result showed that at a particular temperature and strain rate fields, Al alloy 6082-T6 had low flow stress value as compared to AA2024-T4 [11].

Piccini and Svoboda studied the effect of tool penetration depth and sheet positions of the weld joint of dissimilar AA5052–AA6063 Al alloy during friction stir spot welding. The authors found that plastic flow was influenced by tool penetration

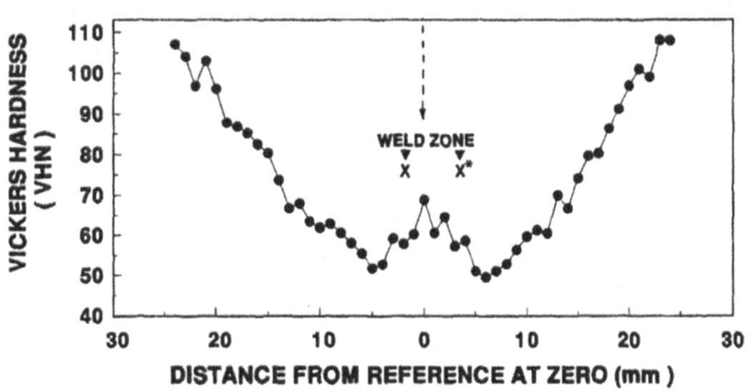

Fig. 1.2 Vickers micro hardness traverse through the section [8]

depth and sheet positions. Their results showed that the effective welded width was increased and hook height decreased with the increase of TPD, when AA6063 alloy was in upper sheet position. Furthermore, the effective welded width was found lower along with a higher hook height, when AA5052 was in upper sheet positions [12].

Cao and Jahazi investigated the effect of probe length and rotational speed on lap joint quality on 2 mm AZ31B Mg alloy using friction stir welding. They concluded that as the rotational speed increased there was an increase in tensile shear load, and a further increase in rotational speed resulted in a decrease in tensile shear load. They also found that the grain advancement appeared in the stir zone at a leading rate of small welding speed as compared to a base metal [13]. Padmanaban and Balasubramanian made an attempt to select the tool pin profile, tool materials and tool shoulder diameter during FSW of AZ31B Mg alloy. They selected four differently shaped pin profile and five types of tool materials to fabricate the weld joint along with 15, 18 and 21 mm tool shoulder diameter. They observed that high carbon steel tool with the threaded pin profile and 18 mm shoulder diameter provided superior tensile properties of joint as compared to other parameters [14]. Kush and Mehta conducted an experiment on dissimilar copper and aluminum workpiece materials using friction stir welding. The authors used a different design of tool pin profile, like cylindrical, square, triangular and hexagonal. They observed that polygonal pin design provided a large and irregular detachment of copper particles from the base material, while in the case of triangular tool pin profile they noticed maximum irregular and large copper particles in the weld zone. They also reported that an increase in the polygonal tool edge decreases the defects, and cylindrical tool pin profile had defect-free macro joint [15].

1.3 Review Related to Optimization Studies of FSW

Process parameter optimization has a significant aspect in the improvement of weld joint properties. It helps in the identification and determination of the appropriate value of process parameters at which the maximum and desired output can be achieved. This section deals with the various optimization techniques used to optimize the FSW process parameters carried out by the researchers.

Farzadi et al. used the response surface methodology to optimize friction stir welding operational parameters for the ultimate tensile strength of AA7075 Al weld joints. A conclusion was drawn that the WS has a more substantial contribution to the ultimate tensile strength, as shown in Fig. 1.3. UTS first declined and then increased with the escalation in WS. They found 513 MPa UTS with 94% joint efficiency under optimum welding condition of 514 rpm rotational speed, 95 mm/min travel speed whereas the tool shoulder and pin diameter were 16.1 and 6 mm, respectively [16]. Shanavas and Dhas used RSM and ANOVA to optimize the FSW parameters of AA5052-S32 Al alloy. They found that the welding parameters, like WS, tilt angle, pin profile and RS, play an extensive role to define the weld joint qualities. Their optimization results revealed that at 600 rpm rotational speed, 65 mm/min WS and

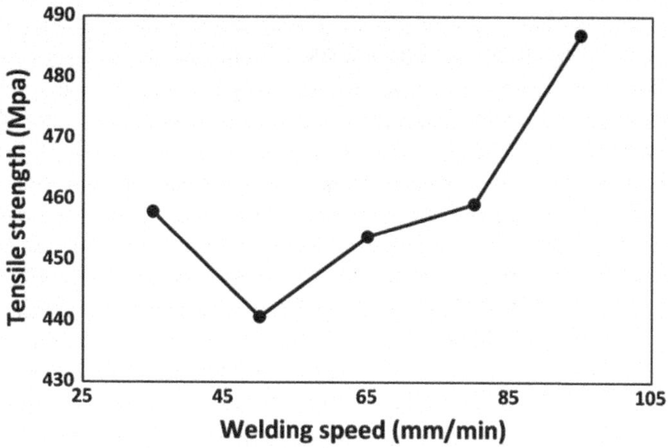

Fig. 1.3 Response of WS on the UTS of weld joints [16]

1.5° tilt angle, the weld joint exhibited superior tensile properties with 93.51% joint as compared to other joints. Moreover, a defect-free weld joint in and around the stir zone was observed during macro and microstructure study of the weld [17]. Kumar et al. analyzed the UTS and percentage elongation of dissimilar AA5083-H111 and AA6082-T6 Al alloy weld joint during FSW and optimized the input parameters with RSM. The input parameters were tool pin profile, RS, WS and axial force. They observed higher value of UTS and percentage elongation of 234 MPa and 12.07%, respectively, at 1000 rpm RS, 90 mm/min WS and 7.8 kN axial load along with tapered square pin profile [18].

Laxminarayanan and Balasubramanian applied Taguchi parametric design and optimization approach technique to determine the effect of process parameters on tensile strength of RDE40 Al alloy weld joint during FSW. A non-consumable tool made of high carbon steel with threaded cylindrical pin profile was used to fabricate the joints. Their optimum values were 1400 rpm RS, 45 mm/min travel speed and 6 kN axial force, which contributed 41, 33 and 21% to tensile strength of weld joints, respectively [19]. Koilraj et al. performed Taguchi method using ANOVA to optimize the process parameters, such as rotational speed, tool speed, tool geometry and the ratio between tool shoulder and pin diameter with respect to tensile strength of weld joint during FSW of dissimilar Al alloy.

AA2219 with AA5083: They found that the tool pin geometry and welding speed played a significant role in deciding the joint soundness (JS), while diametrical ratio of tool to pin diameter was an utmost dominating element for JS and cylindrical threaded pin tool profile was obtained to be the best as compared to other tool pin profiles. Their optimization results were 700 rpm RS, 15 mm/min travel speed, 3 D/d ratio [20]. Elanchezhian et al. have optimized the process parameters of friction stir welded AA8011-6062 Al alloy. They used Taguchi method for condition optimization and ANOVA for evaluating their experimental results. They found WS has negligible effect on the tractile tenacity of the welded joint and was the 153 MPa maximum tenacity at optimal condition of 1400 rpm RS, 75 mm/min WS and 7 kN axial force [21]. Taguchi's grey-based approach has been used by Kesharwani et al. for the multi-objective optimization of FSW parameters of 2 mm thin dissimilar Al sheet AA5052-H32 and AA574-H22 that affect the weld joint quality. They used L9 orthogonal array for their design of experiment. Their optimized parameters to fabricate the joint were 1800 rpm rotational speed, 50 mm per minute travel speed and 20 mm tool shoulder diameter with square pin geometry. They also found that maximum weld strength was 175 MPa and 13.854% maximum elongation at optimum values of welding parameters [22]. Taguchi L9 array was considered by Sachin et al. for the optimization of operational parameters like RS, WS and tool tilt angle during FSW of aluminum matrix composites. They used ANOVA and found that RS was the most influencing parameters as compared to WS for determining tensile strength and hardness of weld joint. Furthermore, they concluded that RS of 1200 rpm and 3° tilt angles were the optimal parameters for UTS and hardness, while optimum WS was 60 mm/min for UTS and 80 mm/min for hardness [23].

Khansari et al. optimized the number of experimental test and mechanical properties of welded FSW of 3 mm thick 2024 Al alloy using Mamdani-type fuzzy logic. They found that high tool rotational speed or low tool forward speeds caused the alter material to plastic phase and produced defects in weld zone. Their optimal value of rotational and forward speed was 1000 rpm and 90 mm/min, respectively [24]. Ashok et al. developed a regression model for optimizing the FSW parameters of Stir cast AA6061-T6/AA1NP composite materials to predict ultimate tensile strength and percentage elongation. They observed that UTS and percentage elongation of joint was 1% greater at the optimum condition, but 9.1% lesser than joint of work material as per the design matrix [25]. Roshan et al. investigated an experimental work for optimization of FSW process parameters of AA7075 Al alloy to achieve desirable mechanical properties such as tensile strength, yield strength and hardness of weld joint by using a simulated annealing algorithm. They utilized four factors and five levels of central composite design for minimizing the number of experimental observations. Their single response and multi-response optimization results showed that at 1400 rpm RS, 1.75 mm/s travel speed and 7.5 kN axial force desirable mechanical properties of weld joint were achieved. Also, their confirmation tests and its verifications showed that the simulated annealing algorithm was an appropriate tool for optimization of FSW process parameters [26]. Ajith et al. used particle swarm optimization method to optimize the friction welding process parameters of UNS S32205 duplex stainless steel. They found the tensile strength to be 827.17 MPa and the maximum hardness of weld joint at 322 HV, which were higher than the tensile strength and hardness of the base material, and the nature of fracture occurred was ductile, as displayed in Fig. 1.4. Their optimal result was 105 MPa frictional pressure, 180 MPa upsetting pressure and 2000 rpm rotational speed [27].

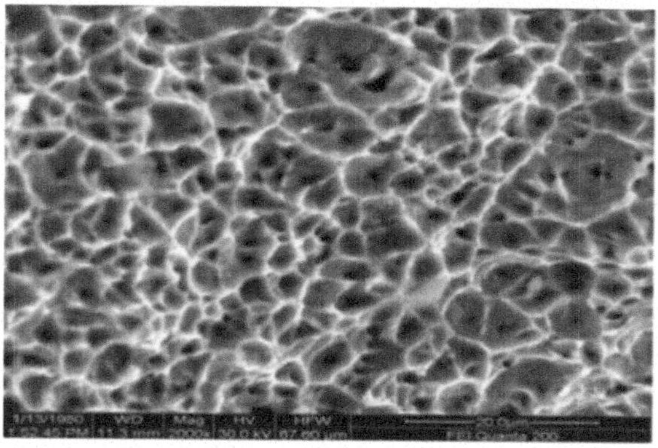

Fig. 1.4 Fractured surface of tensile tested weld sample [27]

1.4 Conclusion

Joining of soft materials like aluminum and copper with high mechanical and microstructural properties is not achieved by fusion welding processes because of the generation of blow holes, brittle inter-metallic compound, porosity and solidification cracking. In order to minimize this, friction stir welding technique is more relevant as the melting of base metal does not take place in this process. FSW is also used to join hard materials like steel and titanium. Following are the outcomes of present work:

- The weld joint properties are mainly affected by the RS, WS and tool geometries. Thus tensile strength, hardness and percentage elongation of weld joint can be improved by varying these operational parameters.
- Discussion on optimization techniques has revealed that at optimum parameters the weld joint possesses good mechanical and microstructural properties.
- Most of the researchers used the Taguchi optimization method, except a few studies that were carried out on the particle swarm optimization method. Therefore, a lot of scopes are available for future work in the area of optimization of FSW process.

References

1. Baffari, D., Buffa, G., Campanella, D., Fratini, L., Micari, F.: Friction based solid state welding techniques for transportation industry applications. Procedia CIRP **18**, 162–167 (2014)
2. Kassim, M.H.: Comparative study between friction stir welding and metal inert gas welding of 2024 T4 aluminum alloy. ARPN J. Eng. Appl. Sci. **6**(11), 36–40 (2011)
3. Thirunavukarasu, G., Chatterjee, S., Kundu, S.: Scope for improved properties of dissimilar joints of ferrous and non-ferrous metals. Trans. Nonferrous Metal. Soc. **27**(7), 1517–1529 (2017)
4. Magalhaes, V.M., Leitao, C., Rodrigues, D.M.: Friction stir welding industrialization and research status. Sci. Technol. Weld. Join. **22**, 1 (2017)

5. Madhavi, B., Jeevan, J., Teja, M.K.: Optimization study of friction stir welding process parameters and nugget properties of aluminium alloy. In: Proceedings of the International Conference on Structural Integrity (ICONS), pp. 4–7. Kalpakkam, India (2014)
6. Nandan, R., Roy, T.D., Bhadeshia, H.K.: Recent advances in friction stir welding-process, weldment structure and properties. Prog. Mater. Sci. **53**, 980–1023 (2008)
7. Singh, M.K., Porwal, R.K.: An overview on solid state welding process. In: National Seminar on Advances in Materials, Manufacturing and Renewable Energy Systems, IJTRA, Lucknow (2018)
8. Liu, G., Murr, L.E., Niou, C.S., McClure, J.C., et al.: Microstructural aspects of the friction-stir welding of 6061-T6 aluminum. Scr. Mater. **37**(3), 355–361 (1997)
9. Saeid, T., Abdollah-zadeh, A., Sazgari, B.: Weldability and mechanical properties of dissimilar aluminum-copper lap joints made by friction stir welding. J. Alloys Compd. **490**, 652–655 (2010)
10. Yabuuchi, K., Tsuda, N., Kimura, A., Morisada, Y., et al.: Effects of tool rotation speed on the mechanical properties and microstructure of friction stir welded ODS steel. Mater. Sci. Eng. A **595**, 291–296 (2014)
11. Fratini, L., Buffa, G., Shivpuri, R.: Influence of material characteristics on plasto mechanics of the friction stir welding process T-joints. Mater. Des. **30**, 2435–2445 (2009)
12. Piccini, J.M., Svoboda, H.G.: Effect of the tool penetration depth in friction stir spot welding of dissimilar aluminum alloys. In: International Congress of Science and Technology of Metallurgy and Materials, SAMCONAMET (2013)
13. Cao, X., Jahazi, M.: Effect of tool rotational speed and probe length on lap joint quality of a friction stir welded magnesium alloy. Mater. Des. **32**, 1–11 (2011)
14. Padmanaban, G., Balasubramanian, V.: Selection of friction stir welding tool pin profile, shoulder diameter and material for joining AZ31B magnesium alloy-an experimental approach. Mater. Des. **30**, 2647–2656 (2009)
15. Kush, P.M., Badheka, V.J.: Effects of tool pin design on formation of defects in dissimilar friction stir welding. In: 3rd International Conference on Innovation and Mechatronics Engineering, ICIAME, pp. 513–518 (2016)
16. Farzadi, A., Bahmani, M., Haghshenas, D.F.: Optimization of operational parameters in friction stir welding of AA7076-T6 aluminum alloy using response surface method. Arab. J. Sci. Eng. **42**(11), 4905–4916 (2017)
17. Shanavas, S., Dhas, J.E.R.: Parametric optimization of friction stir welding parameters of marine grade aluminium alloy using response surface methodology. Trans. Nonferrous Met. SOC. **27**, 2334–2344 (2017)
18. Kumar, H.M.A., Venkata, V.R., Shanmughanathan, S.P., Jacob, J., Mohammed, U.I.: Optimization of dissimilar friction stir welding process parameters of AA5083-H111 and AA6082-T6 by CCD-RSM technique. Adv. Manuf. Process. 49–60
19. Lakshminarayanan, A.K., Balasubramanian, V.: Process parameters optimization for friction stir welding of RDE-40 aluminum alloy using Taguchi technique. Trans. Nonferrous Met. Soc. **18**, 548–554 (2008)
20. Koilraj, M., Sundareswaran, V., Vijayan, S., Rao, S.R.K.: Friction stir welding of dissimilar aluminum alloys AA2219 to AA5083 optimization of process parameters using Taguchi technique. Mater. Des. **42**, 1–7 (2012)
21. Elanchezhian, C., Vijaya, B.R., Venkatesan, P., Sathish, S., et al.: Parameter optimization of friction stir welding of AA8011-6062 using mathematical method. In: 12th Global Congress on Manufacturing and Management (2014)
22. Kesharwani, R.K., Panda, S.K., Pal, S.K.: Multi objective optimization of friction stir welding parameters for joining of two dissimilar thin aluminum sheets. In: 3rd International Conference on Materials Processing and Characterization, pp. 178–187. Hyderabad, India (2014)
23. Sachin, K., Narendranath, S., Chakradhar, D.: Process parameters optimization for FSW of AA6061/SiC/fly ash AMCs using Taguchi technique. Emerg. Mater. Res. 1–8 (2018)
24. Khansari, N.M., Namdar, K., Sarbour, M.H.: Optimization of friction stir welding by fuzzy logic. In: Presented at the 13th Iranian Conference of Fuzzy Systems, Tehran, Iran (2013)

25. Kumar, B.A., Murugan, N.: Optimization of friction stir welding process parameters to maximize tensile strength of stir cast AA6061-T6/AlNp composite. Mater. Des. **57**, 383–393 (2014)
26. Babajanzade, R.S., Behboodi, J.M., Teimouri, R., Ahmadi, G.A., et al.: Optimization of friction stir welding process of AA7075 aluminum alloy to achieve desirable mechanical properties using ANFIS models and simulated annealing algorithm. Int. J. Adv. Manuf. Technol. **69**(5–8), 1803–1818 (2013)
27. Ajith, P.M., Birendra, K.B., Sathiya, P., Aravindan, S.: Multi objective optimization of friction welding of UNS S32205 duplex stainless steel. Def. Technol. **11**, 157–165 (2015)

Chapter 2
Experimental and Numerical Procedure for Studying Effect of Ultrasonic Spot Weld Parameters on Metal Joints for Electronic Components

N. Mohan Raj, L. A. Kumaraswamidhas, S. Arungalai Vendan, K. A. Ramesh Kumar, Liang Gao, Surinder Singh, Harpreet Singh, Xiaodong Niu and Akhil Garg

Abstract Electrical and automobile industries have high demands for an effective joining of dissimilar materials because of their enormous use. In the present research, numerical and experimental investigations have been conducted to study the effect of parameters of ultrasonic spot weld on metal joints for electronic components. Copper and brass alloy sheets have been welded using ultrasonic spot welding in the lap configuration. Regression analysis in combination with analysis of variance has been applied to formulate the models. Based on the optimum condition determined for the input parameters, mechanical and microstructural analysis is performed using tensile testing and SEM analysis, respectively. Finite-element modelling (FEM) simulation is conducted for the temperature near heat-affected zone (HAZ) and fusion

N. Mohan Raj · X. Niu · A. Garg (✉)
Intelligent Manufacturing Key Laboratory of Ministry of Education, Shantou University, Guangdong, China
e-mail: akhil@stu.edu.cn

N. Mohan Raj
Department of Mechanical Engineering, Sri Krishna College of Technology, Coimbatore, India

N. Mohan Raj · L. A. Kumaraswamidhas
Department of Mining Machinery Engineering, Indian Institute of Technology (ISM), Dhanbad, Jharkhand, India

S. Arungalai Vendan
School of Electrical Engineering, VIT University, Vellore, India

K. A. Ramesh Kumar
Department of Energy Studies, Periyar University, Salem, India

L. Gao
State Key Lab of Digital Manufacturing Equipment & Technology, School of Mechanical Science and Engineering, Huazhong University of Science and Technology, Wuhan, China

S. Singh · H. Singh
Department of Mechanical Engineering, Indian Institute of Technology Ropar, Rupnagar 140001, Punjab, India

© Springer Nature Singapore Pte Ltd. 2020
I. Singh et al. (eds.), *Trends in Manufacturing Processes*, Lecture Notes on Multidisciplinary Industrial Engineering,
https://doi.org/10.1007/978-981-32-9099-0_2

zone (FZ). The results obtained from experiments are compared with the simulation results obtained from FEM modelling. Slight variations in tensile properties of the weldments that are free from defects have been observed with increase in welding time and frequency. A close agreement between the experimental and simulated values of FEM has been observed, which establishes the validity of reasonable accuracy of the model.

Keywords Electronic components · Ultrasonic metal welding · Brass–copper alloy · FEM · Design of experiments

2.1 Introduction

A complete and defect-free joint between the dissimilar materials has enormous applications in automobile and electrical industries [1]. The extensively used materials for these industrial applications are mainly copper and brass due to their high electrical and thermal conductivity. In addition to the conductivity, these materials possess higher tensile strength and formability. Ultrasonic spot welding (USW) provides the local spots joining of materials in a shot span of time by dissipating a very lesser heat and by retaining the work-piece properties [2, 3]. In various researches, effects of welding parameters (process parameters) on the properties (mechanical and metallurgical) of weldments have been investigated for joining of dissimilar materials by using USW. Decline in weld strength has been observed with an increase in magnitude of the clamping force, which resulted in diminish in plastic deformation near the interface [3–8]. The investigations showed that better welding–brazing joints can be obtained because of excellent wettability. Also, a gradual increment in interface layer thickness and tensile strength has been observed initially, with increase in welding heat and decrement after that [9]. Further, the relation between energy density and weld pressure has been proposed for joining aluminium alloys by USW [10]. In addition to these experimental studies, various algorithms have been used by researchers to optimize the effect of input parameters on the properties (mechanical and metallurgical) of weldments [11–13].

Various researches have been performed to join copper and its alloys by using friction stir welding (FSW) and resistance spot welding (RSW). It has been observed that the process requires pre-surface preparations, high-stability electrodes, exact alignment, with high current for initiating and driving the process [14–16]. From the above discussion, it can be concluded that the intra- and inter-dependencies of ultrasonic weldment properties (mechanical and metallurgical) on input parameters of dissimilar materials require further investigations.

Therefore, in the present work, dissimilar materials, namely brass and copper, have been welded using USW. Input parameters, namely weld time, pressure, and amplitude, have been varied and their effect on quality of weld joints has been investigated by using SEM, ANSYS, and shear test analyses.

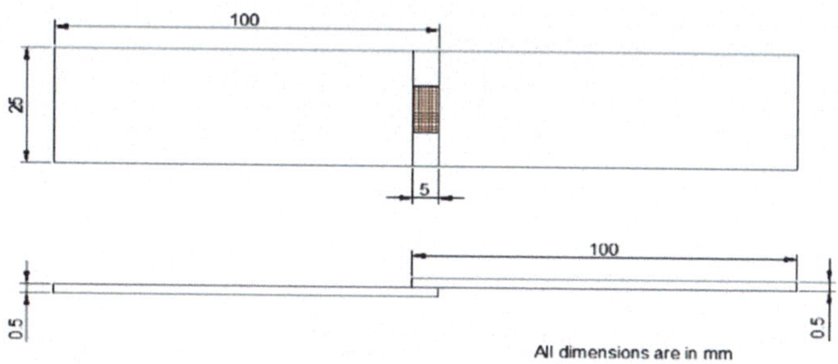

Fig. 2.1 Schematic of weld specimen highlighting the area of weldment

2.2 Methodology

2.2.1 Experimental Setup

During the investigation, the welding has been done between two dissimilar materials, namely, Cu–Ag alloy and brass (Cu–Zn30) of 0.5 mm thin sheets. Chemical composition and mechanical properties of the Cu–Ag alloy and brass have been taken from [16, 17]. All the materials were polished along the rolling direction of sheets by using emery papers to remove oxide layer. Surface roughness measured after polishing was 1.2 and 0.72 μm for Cu–Ag alloy and brass sheets, respectively. The size of each sheet was 100 × 25 mm with an overlap of 5 mm, as represented in Fig. 2.1. Brass was placed below the copper sheets due to its higher hardness as compared to Cu–Ag alloy. Ethanol was used on the brass specimen for lubrication and the gap gets reduced due to capillary effect. Therefore, the coefficient of friction of copper alloys was highly reduced during the welding process. The in terms of T-peel and tensile shear have been evaluated using universal testing machine (UTM) having maximum load capacity of 50 kN. Instron1195 was used to evaluate mechanical strength of the joint. T-peel and tensile shear tests have been performed on UTM by using load cell with capacity of 20 kN.

USW of all the specimens has been done along the lap joint surface, at a vibration frequency of 10–30 kHz and rated power of 2000 W. The anvil has been provided with the patterns similar to the surface for preventing slippage by presuming adequate relative motion between the sheets. In the experimental investigations, temperature of welding zone has been measured during process by placing a K-type thermocouple at the welding zone.

2.2.2 Design of Experiments

The effect of three input variables, namely, vibrational amplitude, welding time, and pressure on the weld quality has been investigated. The values of input parameters

Table 2.1 Process parameters with their levels

Parameters	Level 1	Level 2	Level 3
Applied pressure (bar)	4	4.5	5
Weld time (s)	2	2.5	3
Amplitude (μm)	28	42	57

Table 2.2 L9 orthogonal array setup as per the input variables

Ex. no.	Pressure (bar)	Weld time (s)	Amplitude (μm)	Weld strength Experiments (N/mm^2)
1	4.0	2.0	28	18.21
2	4.0	2.5	42	18.66
3	4.0	3.0	57	18.95
4	4.5	2.0	42	19.74
5	4.5	2.5	57	19.89
6	4.5	3.0	28	19.32
7	5.0	2.0	57	20.47
8	5.0	2.5	28	19.98
9	5.0	3.0	42	20.26

as per their level distribution have been provided in Table 2.1. Orthogonal array of Taguchi's L9 has been used for planning experimentation and weld elements analysis, and a process parametric window has been created, as represented in Table 2.2. At the optimum condition determined for the input parameters, mechanical and microstructural analysis has been performed by using tensile testing and SEM analysis, respectively. Simulations have been performed for temperature near HAZ and FZ. The results obtained from experiments have been compared with the simulation results obtained from FEM modelling.

An impact of input variables on the response of interest (weld strength) has been optimized which comes out to be a main objective of the RSM method in the present study. For forecasting the better welding outputs, second-order mathematical models have been created as done in [18]. Regression analysis was also performed as done by [17–19].

2.3 Finite-Element Modelling (FEM)

The geometry of the finite-element model used to simulate heat conduction has been represented in Fig. 2.2 [20]. To save computation time, the model has been constructed axisymmetrically along the circular weld sonotrode. Furthermore, any

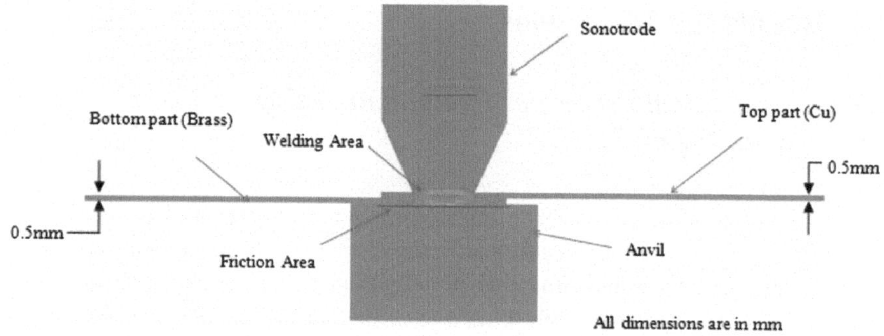

Fig. 2.2 Finite-element analysis heat conduction model for USW of dissimilar materials

heat generated at the top of the sonotrode during the first phase of the welding has been neglected.

The finite-element analysis has been performed over a spot-welded specimen. It uses the mechanical properties of the welded specimen to assess the tensile strength. This has been done by validating the model with data obtained from the shear test. The test joints of the test specimen have been modelled to simulate peak strength under various welding conditions. The adopted values of mechanical parameters for brass and copper in this simulation model have been represented in Table 2.1 and Fig. 2.1. The boundary conditions of FE models for each manufactured overlap test coupon have been set using experimental trial data. Except for a variation in thickness, all work-piece coupons exhibited some symmetry. In the isotopic material model, all spot welding models have a rectangular area comprising 42 elements with a maximal length of 10 mm. A restriction on all degrees of freedom has been placed on the spot welding motion to set the contact condition. The mesh is then overlaid on the weld coupons and has been modelled using solid elements. The normal and tangential interactions between the contact surfaces have been defined by the contact between the spot area and the overlapped work specimens. The same boundary conditions have been considered for the modelling of the welded joints to simulate grip conditions. It becomes necessary to restrict all degrees of freedom, which has been done using an encased boundary condition at one end of the specimen model. This greatly reduces the transverse movement of the load. Also, rotational motion has been neglected at the loading ends of the joints. General contact between metal and metal surfaces has been defined in the friction formulation.

2.4 Results and Discussions

2.4.1 *Mechanical Properties Evaluation of Copper/Brass Weld Region*

The graph between the weld energy and maximum tensile strength has been represented in Fig. 2.3a. A load proportional to the formation of plastic strain has been applied continuously without ultrasonic vibrations until adequate plastic deformation of the material has been achieved. It has been observed that the plastic deformation seen at the weld interface is larger than that at the substrate. The deformation at this interface is due to volume and surface effects. Furthermore, the inter-atomic mechanical interlocking is observed as the primary bonding mechanism. This results in an inter-atomic gap, causing the bonding between surface atoms [21].

The highest energy absorption occurred at level 1, compared with other levels of welding parameters. This absorption depends on the micro-bond formation at the site of welding. Level 3 absorption patterns are similar to level 1, while level 2 requires

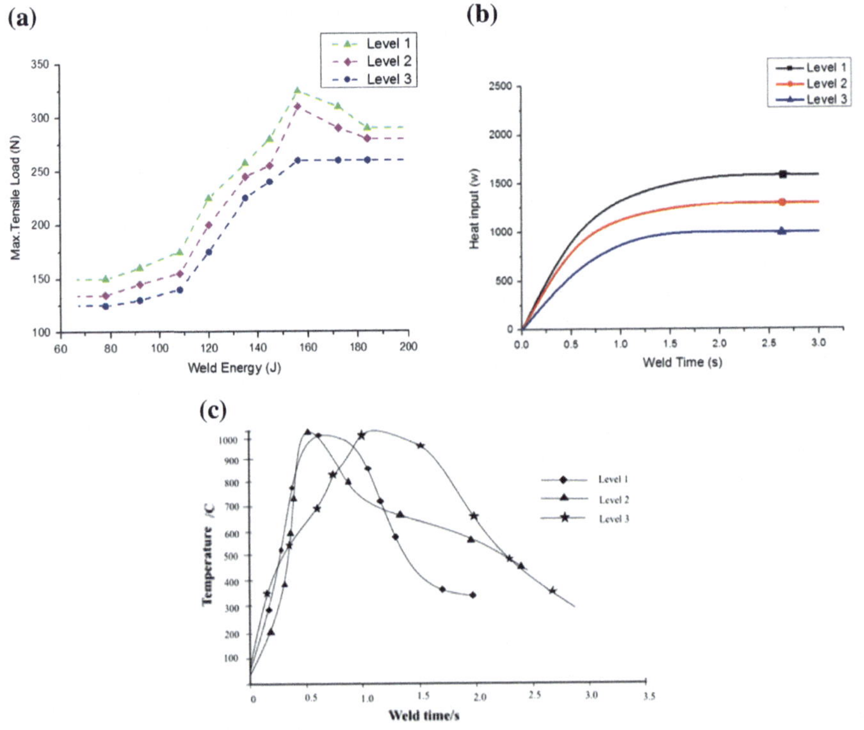

Fig. 2.3 **a** Maximum tensile load versus weld energy; **b** heat fluxes variation; **c** temperature versus weld time behaviour, of different input levels of USW of dissimilar materials at the beginning of a weld cycle

less absorption energy during welding. Levels 1 and 3 display better bonding and weld strength. A careful selection of process parameters can result in welds that maximize tensile strength and hardness. However, a recrystallized microstructure emerges at the weld interface as the weld energy is increased. This is due to a rise in temperature, primarily because of ultrasonic vibrations at the site. The bonding between materials during the welding process is not uniform, which reduces the shear strain.

The different levels of heat flux and weld time variations have been represented in Fig. 2.3b. Linear increase of heat input has been observed during the initial stage of welding for all three levels and becomes content after some time. The apparent power has been measured as an elastic energy stored in the system during welding. Very little apparent power was detected in the parts of the welding system, indicating very small amounts of elastic strain energy.

The variations in the weld zone temperature over welding time have been represented in Fig. 2.3c for various levels. The typical thermal profile at the centre of the joint has been shown for a weld time of 3.0 s. Depending on the magnitude of the temperature, the ultrasonic spot welding thermal cycle for copper and brass has been classified into rising, holding, and cooling temperature phases. The thermal distribution is a major influence on all levels of parameters, which has been measured in a real-time data acquisition system using a thermocouple. It has been observed that a welding time <2 s is not high enough to cause interfacial inter-molecular diffusion [22]. The degree of inter-molecular diffusion between materials has been referred to as the quality of weldments. A significant temperature rise has been observed in level 3, thus requiring more time to create a good joint between brass and copper. It has also been observed that higher temperatures occur only at the edges of the sonotrode, but not in the welding zone. This occurs due to small plastic deformation which occurred at the top of the specimen. It must be emphasized that these deformations do not practically appear in a real weld. These deformations help us conclude that external deformations produce more heat in the weld zone than internal deformations.

2.4.2 Scanning Electron Microscope (SEM)

The SEM images of cross-sectioned samples of poor, good, and over welded joints have been represented in Fig. 2.4. In the under-welded samples (Fig. 2.4a), gaps have been formed between the two materials, thereby preventing the formation of micro-bonds. These can be attributed to the low input parameter values which prevent the development of sufficient friction between the mating surfaces. Therefore, adequate heat generation is necessary for inter-atomic diffusion, whereas the weld strength is mainly attributed to mechanical interlocking. However, interface microstructures have been observed tightly packed and show good micro-bonding under good welding conditions, which is due to the level 1 parameter input settings. Furthermore, bond density is high in a good weld (Fig. 2.4b). These can be attributed to the equate input

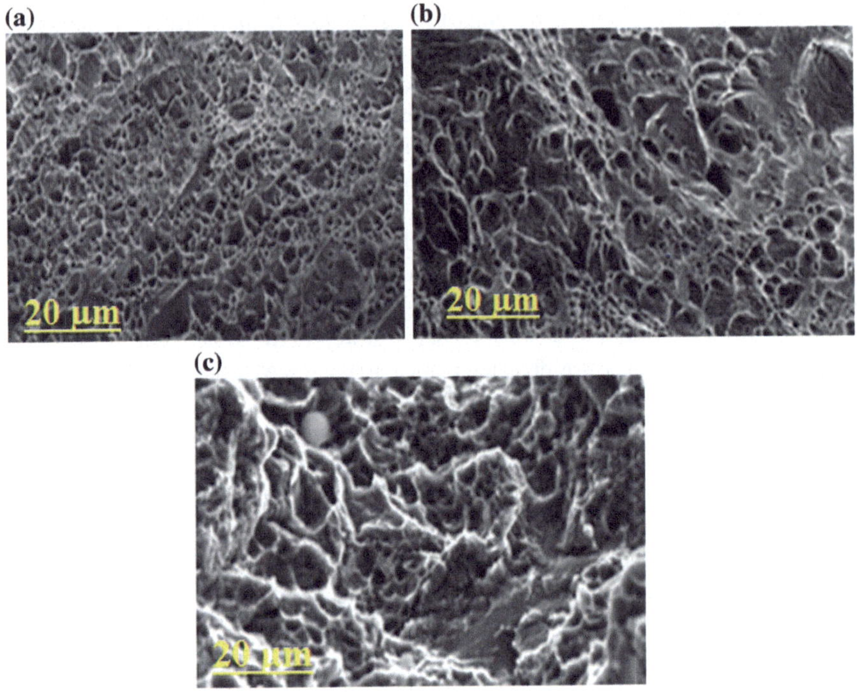

Fig. 2.4 SEM images for different welding parameters **a** under-weld, **b** good weld, and **c** over-weld

parameter values which provides the development of sufficient friction between the mating surfaces. The amount of heat produced at the interfaces directly scales with the values of the input parameters. The resulting inter-atomic diffusion causes an effective bond between the two sheets, classifying it as the good weld condition (Fig. 2.4b).

However, a lack of uniformity in the copper/brass materials has been observed in the over-weld conditions (Fig. 2.4c). A subsequent increase in temperature induces greater ductility in the two sheets. This is accompanied by extreme distortion or extrusion in the top sheet due to the sonotrode tip. This process is referred to as sticking. This study considers copper as the top work-piece. The application of high weld pressure, time, or amplitude may lead to deformation. As copper is more ductile than brass, it may break apart into smaller segments. These conditions may also lead to cracking failures at the edges of the weld.

From ANOVA analysis it has been observed that each level of factor elicits different responses. High-level factors provide responses with means higher than their corresponding lower levels. It has also been seen that the effect of weld pressure as a factor dominates other process parameters. Weld strength has been represented in Table 2.2.

2.4.3 Heat Conduction from the Copper/Brass Weld Zone

The finite-element (FE) modelling of the welded joint has been summarized in Fig. 2.5. The simulated finite-element model used to evaluate the lap-shear strength of the ultrasonic spot-welded coupons has been represented in Fig. 2.5a, with the interfacing surfaces, boundary conditions, and corresponding loading (mashing). The engineering stress and strain values used to explain the properties of the material in the FE simulation has been extracted from prior load displacement data from the tensile test machine. The temperature distribution of the welded model has been represented in Fig. 2.5b. Temperature varied from maximum (1086 °C) at the weld zone towards the minimum at the end of plates (53 °C). Thermal strain distribution (produced due to the heat input by welding process) has been represented in Fig. 2.5c. Maximum strains have been observed near the weld zone and get disappeared after few millimetres distance from the zone. The deformed, non-deformed, and contour patterns of the lap-shear ultrasonic (US) welded coupons have been represented in Fig. 2.5e, and the von Mises displacement and associated stresses of the welded samples are given in Fig. 2.5d. Thus, the complete load–displacement curve for the spot welds has been obtained from the FE models. The results of the experiments from de-bonded specimens have also been obtained similarly.

It has been observed that the vibration amplitude (28, 42, and 57 μm) and clamping force (100, 150, and 200 N) have a significant impact on the load–displacement curves. The peak points of the curve so obtained can be easily used to find the maximum load the welded coupons can withstand. There is clearly an acceptable agreement between the magnitude of the peak points of the experimental results and the FE model predictions. This confirms that the mechanical properties of copper and brass used as input data to the FE model have been correctly extracted, and that the material properties used have been defined properly.

Comprehensive details on both experimental and FEM results for the weld and heat-affected regions have been represented in Table 2.3. The profile shows the dispersion of heat flux from the centre towards the corners of the material with HAZ being considerably narrower. The close agreement between the experimental and estimated values establishes the validity of the model.

2.5 Conclusions

Copper and brass alloy sheets can be welded successfully using ultrasonic spot welding in the lap configuration. A careful selection of process parameters can result in welds that maximize tensile strength and hardness. However, a recrystallized microstructure emerges at the weld interface as the weld energy may increase. This may be due to a rise in temperature, primarily because of ultrasonic vibrations at the site. If the bonding between dissimilar materials during the welding process is observed not uniform, comparatively less shear strain may be observed. Exter-

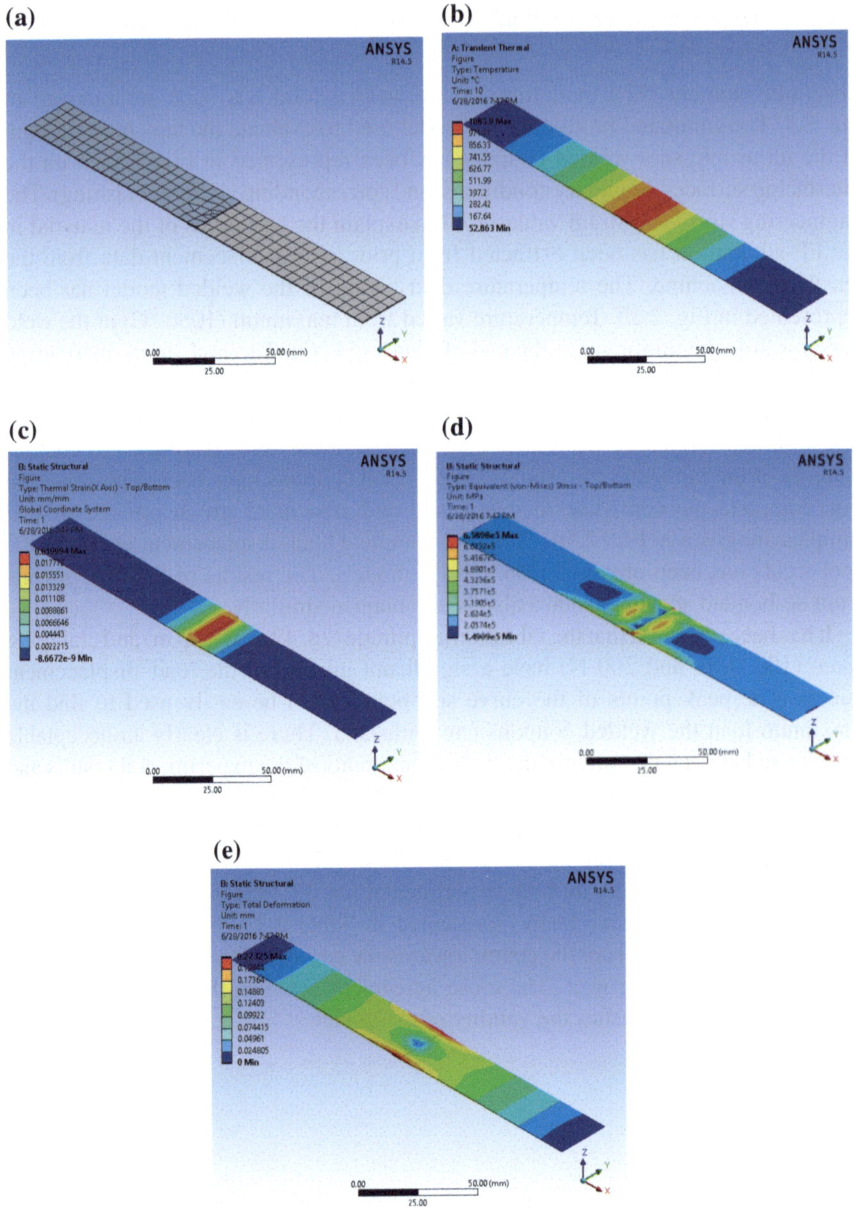

Fig. 2.5 Finite-elemental modelling showing **a** meshing of copper/brass plate, **b** temperature distribution of copper/brass plate, **c** thermal strain of copper/brass plate, **d** equivalent (von Mises) stress copper/brass plate, **e** thermal distribution of copper/brass plate

Table 2.3 Experimental and simulated values obtained from FEM modelling of HAZ and FZ and their percentage error

Parameters/location	Experimental recorded max average temperature (°C)	FE simulation average temperature (°C)	Average error (%)
HAZ	995	976	1.98
FZ	680	626	7.64

nal deformations produce more heat in the weld zone than internal deformations. Adequate heat generation is necessary for inter-atomic diffusion, whereas the weld strength is mainly attributed to mechanical interlocking. The close agreement of the experimental and simulated values of FEM model depicts the high credibility of the model. Future work shall involve the optimization of FEM model for obtaining the approximate values of inputs for desired responses.

References

1. Kerans, W.H.: Welding handbook, 7th edn. Am. Weld. Soc., Miami (1982)
2. Chang, U.I., Frisch, J.: On optimization of some parameters in ultrasonic metal welding. Weld. J. **53**(1), 24 (1974)
3. DeVries, E.: Mechanics and mechanism of ultrasonic metal welding. Ph.D. Dissertation, The Ohio State University (2004)
4. Satpathy, M.P., Bikash, R.M., Shailesh, D., Susanta, K.S.: Modeling and optimization of ultrasonic metal welding on dissimilar sheets using fuzzy based genetic algorithm approach. Eng. Sci. Technol. Int. J. (2015)
5. Watanabe, T., Sakuyama, H., Yanagisawa, A.: Ultrasonic welding between mild steel sheet and Al–Mg alloy sheet. J. Mater. Process. Technol. **34**, 1107–1111 (2009)
6. Sun, Y.F., Xu, N., Fujii, H.: The microstructure and mechanicalproperties of friction stir welded Cu-30Zn brass alloys. Mater Sci. Eng. **589**, 228–234 (2014)
7. Luo, J., Shanshan, L., Wei-Chen, Xiaoling, X.: Friction interface migration of copper alloy and carbon steel dissimilar metal joints in inertia radial friction welding. Mater. Manuf. Process. **31**(3), 275–282 (2016)
8. Zhou, L., Li, Z.Y., Zhao, H.Y., Xie, Y., Huang, Y.X., Feng, J.C.: Microstructure and mechanical properties of Al/brass dissimilar metals TIG welding-brazing joint. Chin. J. Nonferrous Metals **25**(9), 2389–2395 (2015)
9. Erdem, M.: Investigation of structure and mechanical properties of copper-brass plates joined by friction stir welding. Int. J. Adv. Manuf. Technol. **76**(9–12), 1583–1592 (2015)
10. Matsuoka, S., Hisashi, I.: Direct welding of different metals used ultrasonic vibration. J. Mater. Process. Technol. **209**(2), 954–960 (2009)
11. Elangovan, S., Prakasan, K., Jaiganesh, V.: Optimization of ultrasonic welding parameters for copper to brass joints using design of experiments. Int. J. Adv. Manuf. Technol. **51**, 163–171 (2010)
12. Cheng, X., Xiaochun, L.: Investigation of heat generation in ultrasonic metal welding using micro sensor arrays. J. Micromech. Micro-Eng. **17**(2), 273 (2007)
13. Gencsoy, H.T., Admas, J.A., Shigeo, S.: On some fundamental problems in ultrasonic welding of dissimilar metals. Weld. J. **4**, 145s–153s (1967)

14. Chen, Y.-C., Bakavos, D., Gholinia, A., Prangnell, P.B.: HAZ development and accelerated post-weld natural ageing in ultrasonic spot welding aluminium 6111-T4 automotive sheet. Acta Mater. **60**(6), 2816–2828 (2012)
15. Akbari, M.S., Niknejad, S.T.: An investigation on microstructure and mechanical properties of Nd:YAG laser beam weld of copper beryllium alloy. Metall. Mater. Trans. A **40**, 1469–1478 (2009)
16. Prager, M., Dralle, C.W., Gaffoglio, C.J., Wold, K.G., Goldstein, S.H., Waterbury, W.V.: In: Kearns, W.H. (ed.) AWS Welding Handbook Metals and Their Weldability, vol. 4, pp. 278, 7th edn. AWS, New York (1982)
17. Canyurt, O.E., Kim, H.R., Lee, K.Y.: Estimation of laser hybrid welded joint strength by using genetic algorithm approach. Mech. Mater. **40**, 825–831 (2008)
18. Cochran, W.G., Cox, G.M.: Lattice designs. In: Experimental Designs, pp. 396–438. Wiley, New York (1992)
19. Myers, R.H., Montgomery, D.C., Anderson-Cook, C.M.: Response Surface Methodology: Process and Product Optimization Using Designed Experiments. Wiley (2016)
20. De Vries, E.: Mechanics and mechanisms of ultrasonic metal welding. Ph.D. Dissertation, The Ohio State University, (2004)
21. Siddiq, A., Elaheh, G.: Thermo-mechanical analyses of ultrasonic welding process using thermal and acoustic softening effects. Mech. Mater. **40**(12), 982–1000 (2008)
22. Zhang, C.Q., Robson, J.D., Prangnell, P.B.: Dissimilar ultrasonic spot welding of aerospace aluminum alloy AA2139 to titanium alloy TiAl6V4. J. Mater. Process. Technol. **231**, 382–388 (2016)

Chapter 3
Micro-Electrical Discharge Machining for Machining Micro-Holes

M. S. Nagrale and S. A. Mastud

Abstract It is very challenging to machine high aspect ratio and complex-shaped holes with micro-electrical discharge machining process. Micro-hole drilling plays a vital role in the production of various aspect ratio (length/depth) micro-holes. This work presents an attempt to fabricate high aspect ratio micro-holes using micro-EDM process on a brass workpiece with tungsten carbide electrode. In the present study, the effect of different machining and operating parameters, namely, spindle speed, feed, input voltage and capacitance on the performance of micro-EDM process is studied. The performance of micro-EDM process is evaluated on the basis of material removal rate (MRR) and the tool wear rate (TWR). It was found that there is significant effect of capacitance and feed on the MRR. The MRR improved with increasing values of capacitance and feed rates, while the TWR decreases with increasing discharge energy (capacitance and voltage).

Keywords Micro-electro discharge machining (micro-EDM) · Micro-hole · Taguchi method · Tool wear rate · Material removal rate

3.1 Introduction

Micro-machining technologies are gaining importance to produce micro-parts with enhanced product functionality in a limited space. The selection of the micro-machine technique to fabricate micro-holes depends on various parameters, such as aspect ratio for drilling micro-holes, size and shape of workpiece, workpiece material and electrode material [1]. Many researchers are attracted towards research in micro-drilling due to its increasing need of high aspect ratio holes in the aircraft, aerospace and medical applications. Micro-electro discharge machining (micro-EDM) is one of the popular non-conventional machining techniques for drilling accurate and

M. S. Nagrale
Department of Production Engineering, Veermata Jijabai Technological Institute, Mumbai, India

S. A. Mastud (✉)
Department of Mechanical Engineering, Veermata Jijabai Technological Institute, Mumbai, India
e-mail: samastud@vjti.org.in

© Springer Nature Singapore Pte Ltd. 2020

I. Singh et al. (eds.), *Trends in Manufacturing Processes*,
Lecture Notes on Multidisciplinary Industrial Engineering,
https://doi.org/10.1007/978-981-32-9099-0_3

complex-shaped micro-holes. Micro-EDM has a growing popularity due to its advantages, including enhanced precision, low setup cost and significant design freedom [2–6]. Micro-EDM can fabricate a micro-hole using as small as 5 μm diameter holes on any conductive material, irrespective of its melting point and hardness [4]. Apart from these advantages, there are still some limitations in fabrication of micro-EDM to achieve a high aspect ratio holes and tolerances. Jung and Kwon [2] studied the influence of various machining and operating parameters, namely, voltage, capacitance, feed rate and spindle speed on the performance of the micro-EDM process. The Taguchi's grey relational method is used for the analysis of experimental data, and an optimal condition for the high aspect ratio micro-EDM was found. There was an effect of clearance and tool wear on diameter of micro-hole when the same electrode diameter was used. From the grey relation they found that the input voltage and capacitance are the most significant machining parameters. Ferraris et al. [7] proposed an innovative method for the deep-drilling machining using micro-EDM. A tubular electrode coated with low electrically conductive material is used for the reduction of undesirable discharge by suppressing the spark between the tool flank and workpiece. The authors found reduction of tool wear by two-fold with 30% improvement in aspect ratio with coated electrode tool. Lim et al. [3] made three different types of fine electrode for on-machine tool fabrication and assessed their performance in micro-EDM using different parameters. A stationary sacrificial block, rotating sacrificial disk and guided running wire electrodes are used for the preparation of electrode. The authors concluded that a rotating electrode gave best performance for high aspect ratio tool electrode fabrication, and the machining depth is found to be inversely proportional to the feed rate. Yahagi et al. [8] investigated the influence of the high spindle speed on the performance of the micro-drilling EDM. They report that the material removal rate and accuracy can be improved with high rotation of tool electrode; as a result, a much deeper micro-hole with low tool wear rate is possible with high spindle speed.

Numerous researches have reported on the micro-drilling using different electrodes on different workpiece materials [9, 10]. D'urso and Merla [11] demonstrated the influence of different workpiece and electrode material on the micro-EDM process while achieving micro-holes. Four different workpiece materials and three electrode materials with two different shapes were considered to study the performance of micro-EDM. They found an electrode material has a great influence on the TWR than the other parameters. Plaza et al. [12] did an experimental study with the Ti6Al4V alloy material used in aerospace industries and a helical electrode. The influence of helical electrode parameters, namely, helix angle and flute depth, on the process parameters is discussed. Fu et al. [13] did a work to machine a micro-hole with different electrode material under different machining conditions, namely, rough, semi-finish and finish. A relationship between side gap, feed and electrode accuracy was proposed and used to analyse the influence of these parameters on the TWR. The above-mentioned literatures show that there is a lot of work performed to study the influence of various process and machining parameters on the micro-EDM process. But there is a need to investigate and find the optimum process parameters for the setup.

The present study focusses on a detailed investigation of individual and combined effect of process parameters, including input voltage, capacitance, spindle speed and feed rate on a material removal rate and tool wear ratio for a high aspect ratio micro-EDM process. A tungsten carbide electrode was used for fabricating deep holes in brass material. Experiments were planned by using mixed-level Taguchi orthogonal array (OA) design method to analyse the effects of input voltage, capacitance, spindle speed and feed rate on material removal rate (MRR) and tool wear rate (TWR).

3.2 Experimental Setup and Method

3.2.1 Experimental Setup

A Sinergy nano systems Hyper-10 EDM machine is used for the machining of the micro-hole in a set of experiments. Hyper-10 EDM is a table-top multipurpose miniature machine tool that performs multiple machining processes, like micro-turning, micro-milling, micro-grinding and micro-EDM drilling, with high precision. Figure 3.1 shows the conceptual process of micro-EDM to fabricate a high aspect ratio hole. A relaxation-type circuit pulse generator is used to provide a discharge between electrode and workpiece. A hybrid precision original motion controller with original GUI system is installed to control the system with a standard NC code. A computer containing HYPER-10 software is connected to an EDM controller to control the displacement in X, Y and Z direction. The maximum travel range of the machine has its size of $130 \times 75 \times 80\ \text{mm}^3$ with the resolution of 0.1 μm in X, Y and Z direction, respectively, with a 5 micron position accuracy. A BLDC motor of capacity up to 3000 rpm rotational spindle speed is installed in the spindle system to achieve micro-machining. The length and diameter of electrodes and a fabricated micro-hole are measured by using Mitutoyo-MF series measuring microscope with 5 μm accuracy. Rapid I vision measuring machine system is used to process the images and dimensions of the hole.

3.2.2 Machining Electrode Tool and Workpiece

A high melting point tungsten carbide solid rod of length 40 mm and diameter 500 μm is used for machining micro-holes. A commercially available Total FINA ELF EDM 3 dielectric fluid having high flash point, high auto ignition temperature, low pour point and high dielectric strength has been used. This dielectric fluid flushes the minute chips eroded from the workpiece. It serves as a conducting medium when ionized and conveys the spark. It concentrates the energy to a very narrow region. Brass material of 2.5 mm thickness plate is used as a workpiece. The properties of the workpiece, tool electrode and dielectric fluid are summarized in Tables 3.1 and 3.2.

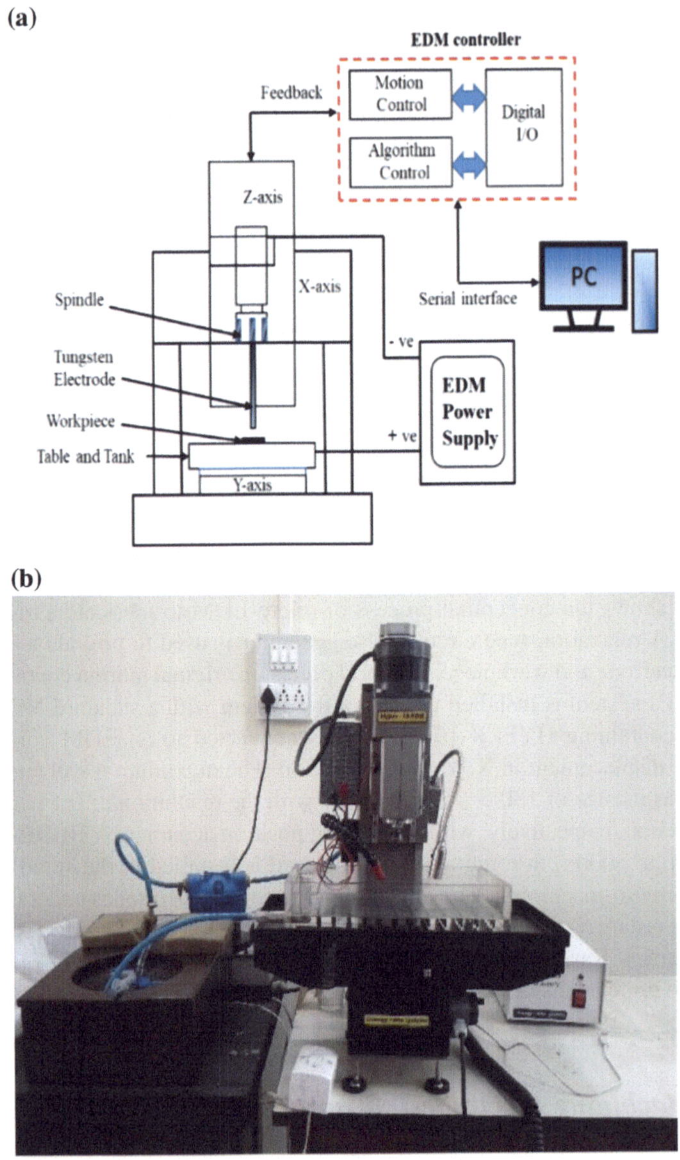

Fig. 3.1 **a** Schematic diagram of experimental setup. **b** Photograph of experimental setup

Table 3.1 Properties of tool and work material

	Workpiece	Electrode tool
Material composition (wt%)	Cu: 60%, Zn: −35% + others	90% W
Density (g/cm^3)	8.47	15
Melting point (°C)	620	2850
Electrical resistivity (Ω cm)	–	65×10^{-6}
Thermal conductivity (W/mK at 25 °C)	109	75
Hardness	60	90

Table 3.2 Properties of the dielectric fluid EDM oil 3 [14]

Volumetric mass at 15 °C (kg m^3)	813
Viscosity at 20 °C (mm^2/S)	7
Flash point (°C)	134
Auto ignition temperature (°C)	243.33
Distillation range IBP/FBP (°C)	277/322

3.2.3 *Experimental Method for Micro-Hole Drilling*

The purpose of the experiments is to fabricate high aspect ratio micro-holes. In EDM process, the discharging happens when the gap between workpiece material and tool electrode is sufficiently close to generate a spark. The z-axis of the machine is used to make a sufficient gap between workpiece material and electrode. The gap is adjusted with the required value of input voltage, as it measures when the spark generates between the gap. For higher MRR and controlled performance, negative electrode polarity is selected. For the experiments, input voltage, capacitance, feed and spindle rotation were selected as process parameters for the performance analysis of micro-EDM process. MRR is the ratio of volume of material removed from part to the machining time, whereas TWR is the ratio of volume of material removed from electrode to the volume of material removed from part. Material removal rate (MRR) and tool wear ratio (TWR) are calculated using the following equations:

$$MRR = \frac{\pi D_m^2 L}{4t} \tag{3.1}$$

where D_m is the mean diameter of the hole ($D_m = \frac{d1+d2}{2}$), and L is the length of the hole.

$$TWR = \frac{D_e^2(L_i - L_f)}{D_m^2 L} \times 100 \quad (3.2)$$

3.2.4 Experimental Design and Analysis

A Taguchi method-based experimentation scheme was used to determine the main effects and statistical significance of various machining parameters to fabricate a high aspect ratio hole with possible minimum diameter. Table 3.3 shows the selected machining parameters for the experimentation. An L18 orthogonal array of Taguchi method was used to perform different trial conditions for experiments with four factors and three levels. Two levels of capacitance and three levels of other parameters were selected to fix the trial conditions. For complete analysis of experimental data, standard analysis of variance (ANOVA) approach for raw data analysis and signal-to-noise (S/N) linked with a loss function for multiple runs were used. The S/N ratio determines the most robust set of operating conditions from variation within results. In the present work, the raw data analysis and S/N data analysis have been performed, and the effect of selected process parameters on the EDM process has been investigated through the main effect plots based on raw data. The optimum condition for each of the quality characteristics has been established through S/N data analysis aided by the raw data analysis. A high value of S/N ratio implies that signal is much higher than the random effect of noise factor. The S/N ratio results may be investigated in three different ways: "smaller the better", "larger the better" and "nominal the best" and are given as follows [15]:

Larger is better,

$$(\text{S/N}) = -10\log(MSD) \quad (3.3)$$

where

Table 3.3 Machining conditions and input levels for EDM drilling

Process parameters	Experimental conditions		
	1	2	3
Capacitance (nF)	3	4	
Voltage (V)	125	150	175
Feed rate (μm/s)	5	10	15
Spindle speed (rpm)	500	1000	1500

$$MSD = \frac{1}{R}\sum_{j=1}^{R}\left(\frac{1}{y_j^2}\right) \tag{3.4}$$

Smaller is better,

$$(\text{S/N}) = -10\log(MSD) \tag{3.5}$$

$$MSD = \frac{1}{R}\sum_{j=1}^{R}(y_j^2) \tag{3.6}$$

Nominal is better,

$$(\text{S/N}) = -10\log(MSD) \tag{3.7}$$

$$MSD = \frac{1}{R}\sum_{j=1}^{R}(y_j - y_o)^2 \tag{3.8}$$

3.3 Results and Discussion

The main purpose of this work is to achieve the highest possible high aspect ratio micro-hole. From the preliminary experiments, a brass plate of 250 μm thickness was selected as a work material. Table 3.4 illustrates the quantitative results of MRR and TWR for different experiments. From the quantitative results, it was found that the highest material removal rate (MRR) and the lowest value of tool wear rate (TWR) are achieved with 3 nF capacitance, 150 V voltage, 15 μm/s feed rate and 1500 rpm spindle speed.

According to Taguchi methodology, the parameters with large values represent better performance for micro-EDM and termed as "larger-the-better", such as material removal rate. On the other hand, the TWR should be minimum for the better economy and high performance. Hence, smaller values of TWR indicate better performance and termed as "smaller-the-better". All the experiments are repeated two times, and an optimal combination of the machining and process parameters was evaluated (Fig. 3.2).

Table 3.4 Experimental values of MRR and TWR for different experiments

Experiment no.	Capacitance (nF)	Voltage (V)	Feed (m/s)	Spindle speed (rpm)	MRR (mm^3/min)	TWR
1	3	125	5	500	0.0433	0.10154
2	3	125	10	1000	0.0765	0.08750
3	3	125	15	1500	0.0757	0.08153
4	3	150	5	500	0.0728	0.10329
5	3	150	10	1000	0.0888	0.13061
6	3	150	15	1500	0.1458	0.04239
7	3	175	5	1000	0.0700	0.07135
8	3	175	10	1500	0.0896	0.10018
9	3	175	15	500	0.1388	0.10827
10	4	125	5	1500	0.0639	0.07922
11	4	125	10	500	0.1046	0.06461
12	4	125	15	1000	0.1333	0.06179
13	4	150	5	1000	0.0495	0.08008
14	4	150	10	1500	0.0862	0.07365
15	4	150	15	500	0.0497	0.12420
16	4	175	5	1500	0.0581	0.17253
17	4	175	10	500	0.1137	0.10564
18	4	175	15	1000	0.0938	0.09924

3.3.1 *Effect of Process Parameters on MRR*

Figure 3.3 depicts the effect of process parameters with S/N ratio curve. The figure confirms that MRR increases with lower capacitance and the peak S/N ratio value is found at lower capacitance of 3 and 4 nF. The S/N ratio is found maximum with highest voltage, and MRR increases with the increment in voltage and feed. At high voltage, the explosive energy is high; hence the discharge spark increases, which results in increased MRR. From the figure, feed versus S/N ratio, it is clear that MRR increases significantly with feed rate of 5–10. After this the S/N ratio value reduces with marginal value for the experiment. The plot also shows that at lower value of spindle speed, the material removal rate was found higher. From the above results it is concluded that voltage and feed are influential parameters on MRR.

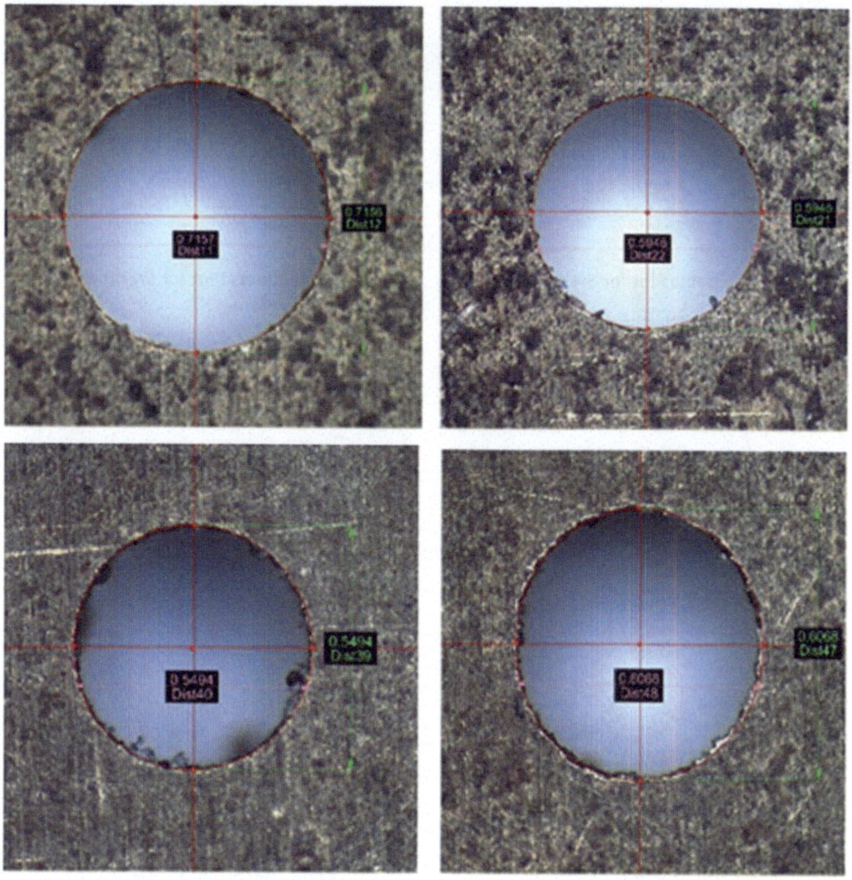

Fig. 3.2 Microscopic view at entry and exit of micro-hole

3.3.2 Effect of Process Parameters on TWR

Figure 3.4 shows the S/N ratio plot for tool wear rate (TWR). As explained in Sect. 3.3, for the TWR, "smaller the better" is considered. There is a significant effect of voltage, spindle speed and feed rate on the TWR than the capacitance. From the plot it is shown that for lower capacitance 4 nF, the S/N ratio is lower; hence it is concluded that higher capacitance value lowers the TWR with reduction in cost of EDM process. The S/N ratio reduces with increase in voltage. On the other hand, with increase in feed rate and spindle speed, the S/N ratio increases but after 1000 rpm spindle speed, the effect is marginal. It is concluded from the above results that there is a significant effect of voltage, feed rate and spindle speed on the tool wear rate. The MRR is found maximum and TWR is found minimum with 3 nF capacitance, 150 V voltage, 15 μm/s feed rate and 1500 rpm spindle speed. 10% increase in the

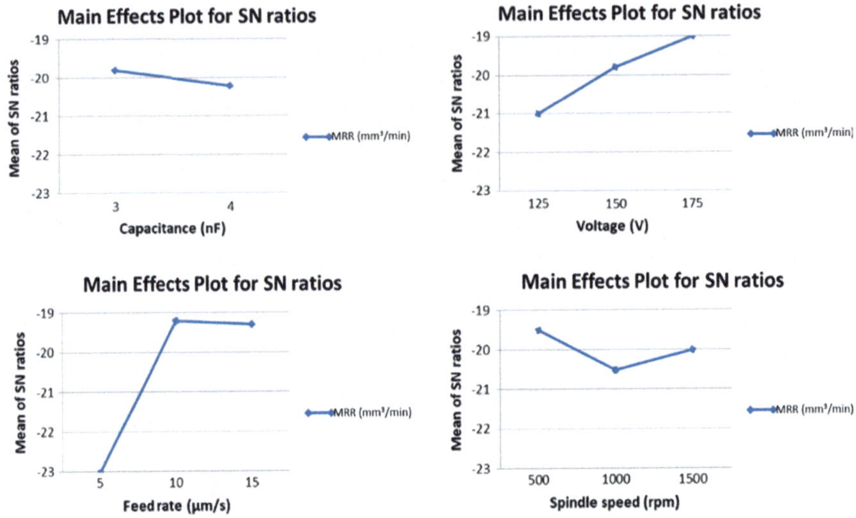

Fig. 3.3 S/N ratio plot of different process parameters for material removal rate

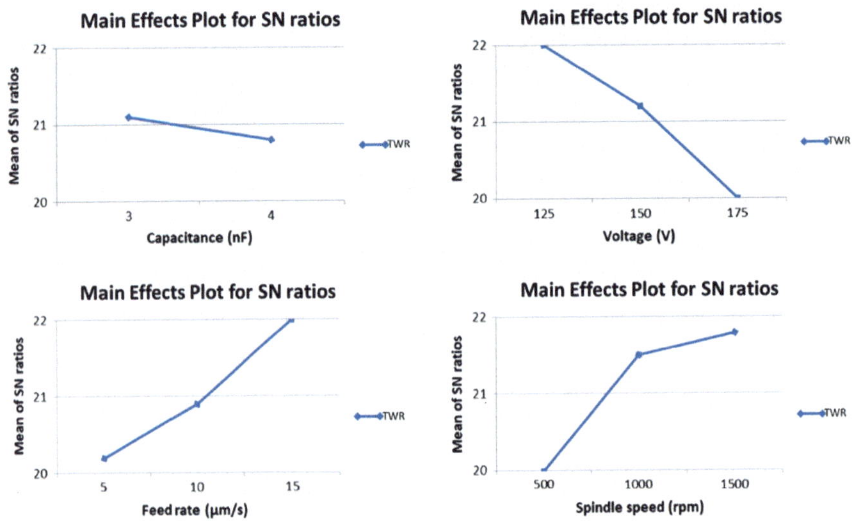

Fig. 3.4 S/N ratio plot of different process parameters for tool wear rate (TWR)

Table 3.5 Analysis of variance for signal to noise ratios

Source	DF	SS	MS	F	P
Voltage	2	21.357	10.6787	1.13	0.360
Capacitance	1	0.301	0.3006	0.03	0.862
Feed rate	2	7.264	3.6320	0.39	0.690
Spindle speed	2	7.632	3.8159	0.41	0.677

MRR is found as the discharge energy increases with the increase in voltage and capacitance.

3.3.3 Analysis of Variance (ANOVA)

The relative significant process parameters are determined with the analysis of variance (ANOVA) approach. The F-values are calculated for MRR and TWR and illustrated in Table 3.5. From Table 3.5, only one parameter voltage has significant effect on MRR. Therefore, the results confirm that voltage is the most significant optimal parameter for EDM process design.

3.4 Conclusion

A different set of experiments is performed to fabricate micro-hole with micro-EDM process. The influence of various machining and process parameters, such as voltage, capacitance, feed rate and spindle speed, on the performance of EDM, like MRR and TWR, has been examined. A Taguchi analysis is performed to attain an optimal machining and process parametric combinations. From the results, the following outcomes are observed during micro-EDM process.

Voltage and feed rate are the most influencing factors on the material removal rate (MRR) and micro-EDM process. Material removal rate increases with increase in voltage and feed. Also, MRR has been found to be increased with lower spindle speed of 500 rpm, but after 500 rpm of spindle speed the effect is marginal. There is very less effect of capacitance on the MRR.

There is a significant effect of voltage, feed rate and spindle speed on the tool wear rate (TWR). A smaller value is found for 4 nF capacitance but the effect is marginal. With an increment in voltage, the S/N ratio achieved smaller value, while the S/N ratio increases with increase in feed rate. Tool wear rate is less with lower spindle speed as the S/N has a smaller value for 500 rpm.

Based on the Taguchi analysis, the F-values are determined from ANOVA, and optimal parametric conditions are achieved for this type of EDM process. From the analysis, it is concluded that for higher MRR and lower TWR, 3 nF capacitance, 150 V voltage, 15 μm/s feed rate and 1500 rpm spindle speed parametric combination is found effective. Voltage has the highest F value. With this combination the highest value of aspect ratio is achieved as 5.

References

1. Yu, Z.Y., Rajurkar, K.P., Shen, H.: High aspect ratio and complex shaped blind micro holes by micro EDM. CIRP Ann. Manuf. Technol. **51**(1), 359–362 (2002)
2. Jung, J.H., Kwon, W.T.: Optimization of EDM process for multiple performance characteristics using Taguchi method and Grey relational analysis. J. Mech. Sci. Technol. **24**(5), 1083–1090 (2010)
3. Lim, H.S., Wong, Y.S., Rahman, M., Lee, M.E.: A study on the machining of high-aspect ratio micro-structures using micro-EDM. J. Mater. Process. Technol. **140**(1–3), 318–325 (2003)
4. Puranik, M.S., Joshi, S.S.: Analysis of accuracy of high-aspect-ratio holes generated using micro-electric discharge machining drilling. Proc. Inst. Mech. Eng. Part B J. Eng. Manuf. **222**(11), 1453–1464 (2008)
5. Yadav, U.S., Yadava, V.: Parametric study on electrical discharge drilling of aerospace nickel alloy. Mater. Manuf. Process. **29**(3), 260–266 (2014)
6. Dewangan, S.K.: Experimental investigation of machining parameters for EDM using U-shaped electrode of AISI P20 tool steel. Department of Mechanical Engineering, National Institute of Technology Rourkela (India) (2010)
7. Ferraris, E., Castiglioni, V., Ceyssens, F., Annoni, M., Lauwers, B., Reynaerts, D.: EDM drilling of ultra-high aspect ratio micro holes with insulated tools. CIRP Ann. **62**(1), 191–194 (2013)
8. Yahagi, Y., Koyano, T., Kunieda, M., Yang, X.: Micro drilling EDM with high rotation speed of tool electrode using the electrostatic induction feeding method. Procedia CIRP **1**, 162–165 (2012)
9. Karthikeyan, G., Ramkumar, J., Dhamodaran, S., Aravindan, S.: Micro electric discharge milling process performance: an experimental investigation. Int. J. Mach. Tools Manuf. **50**(8), 718–727 (2010)
10. Gil, R., Sánchez, J.A., Ortega, N., Plaza, S., Izquierdo, B., Pombo, I.: High-aspect ratio micro-pin manufacturing using inverse slab electrical discharge milling (ISEDM) process. Int. J. Adv. Manuf. Technol. **65**(9–12), 1459–1469 (2013)
11. D'Urso, G., Merla, C.: Workpiece and electrode influence on micro-EDM drilling performance. Precis. Eng. **38**(4), 903–914 (2014)
12. Plaza, S., Sanchez, J.A., Perez, E., Gil, R., Izquierdo, B., Ortega, N., Pombo, I.: Experimental study on micro EDM-drilling of Ti6Al4V using helical electrode. Precis. Eng. **38**(4), 821–827 (2014)
13. Fu, Y., Miyamoto, T., Natsu, W., Zhao, W., Yu, Z.: Study on influence of electrode material on hole drilling in micro-EDM. Procedia CIRP **42**, 516–520 (2016)
14. Jahan, M.P.: Micro-EDM-based multi-process machining of Tungsten carbide. Doctoral dissertation (2009)
15. Phadke, M.S.: Quality Engineering Using Robust Design. Prentice Hall PTR (1995)

Chapter 4
Nd:YVO$_4$ Laser Welding of Two Transparent Polymers in Lap Joint Configuration

Nitesh Kumar, Nikhil Kumar and Asish Bandyopadhyay

Abstract The present research manages through transmission laser welding (TTLW) of polycarbonate and acrylic, both being transparent, have been directed framing a lap joint. A 12-W Nd:YVO$_4$ laser with 1064 nm wavelength has been used for the welding procedure. The effect of welding parameters, to be specific, scanning speed, laser power and laser frequency, has been analysed. The weld quality has been assessed by an ultimate load (UL). Response surface methodology (CCD) has been embraced for the experimental procedure. To correlate the welding parameters with a response, a quadratic model has been created. The analysis of variance method is used to find out the important parameter/s on specific response. A mathematical model has been used to optimize the ideal welding conditions to achieve the improved ultimate load. To validate the applied optimization technique, confirmatory tests have been performed.

Keywords Ultimate load · RSM · ANOVA · Laser welding

4.1 Introduction

The transmission laser welding has been widely studied in the present day because of its many advantages, including precision, very small heat-affected zone due to the focussed application of energy, which results in lower distortion, extensive types of

---

N. Kumar
Mechanical Engineering Department, Brainware Group of Institution-SDET, Kolkata 700124, India
e-mail: nitesh.ju@gmail.com

N. Kumar (✉)
Centre for Precision Engineering, Materials and Manufacturing Research, Institute of Technology, Sligo F91 YW50, Ireland
e-mail: nikhilju2013@gmail.com

A. Bandyopadhyay
Mechanical Engineering Department, Jadavpur University, Kolkata 700032, India
e-mail: asishbanerjee@yahoo.com

© Springer Nature Singapore Pte Ltd. 2020

I. Singh et al. (eds.), *Trends in Manufacturing Processes*,
Lecture Notes on Multidisciplinary Industrial Engineering,
https://doi.org/10.1007/978-981-32-9099-0_4

joint geometry and materials, ability to be automated and so on. Laser welding of transparent plastic materials has been among the fields of growing interest. Transparent polymers like acrylic and polycarbonate have many applications in medical devices, electronics, packaging and so on, due to their favourable properties such as light transmission, lightweight, frame-resistance, high impact strength, UV protection, aesthetic appearance and non-toxic nature, while acrylic is also known for good insulation properties, ease of processing, weather resistance and durability. The components made by welding acrylic and polycarbonate materials may certainly constitute a combination of the properties of both the materials, and hence can be optimized for more versatile applications. The transmission laser welding is done typically, where one part is transparent, and the other is opaque. Application of an additive can achieve TTLW of two transparent parts, and very limited research has been done so far in this area.

In the work of Coelho et al. [1], the study of welding of thin white transparent film of polypropylene and polyethylene of low and high densities at a high speed of 20 m/s using a CO_2 laser has been conducted. Haberstroh et al. [2] examined how the weld seam formation is influenced by carbon black content during the joining of thermoplastics in micro-technology applications. It has been noticed that the penetration depth decreased from 29 to 9 μm, with an increase in the carbon black content between 0.5 and 1.5%. The effect of specific nylon composition factors, such as fibreglass, mineral, impact modifier content and the colour version on the near infra-red transmission has been studied by Bray et al. [3]. Jansson et al. [4] observed that quasi-simultaneous laser welding of plastics is a fast, reliable and flexible welding method to join plastics. It has been found that the polypropylene (PP) and polycarbonate (PC) samples used in quasi-simultaneous laser welding enabled stronger welds, a larger process window and better air-gap bridging capability than contour welding. Laser transmission welding of glass-reinforced nylon-6 in T-joint configuration has been studied by Prabhakaran et al. [5]. Kumar et al. [6] have done multi-objective optimization of the laser transmission welding of thermoplastics (acrylic and polycarbonate) using grey-based Taguchi method. Roth et al. [7] used femtosecond laser to weld transparent plastics with powers between 500 and 2200 mW. Little or almost no information is available on transparent-to-transparent polymer welding using Nd:YVO_4 laser, and much exploration is needed in order to build up a clear insight of TTLW of two transparent polymers in lap joint configuration. Experiments have been carried out with three factors and five levels, and 15 tests have been performed, including one centre point. Single-objective optimization has been used using desirability function analysis to maximize the ultimate load.

4.2 Experimental Setup and Procedure

In the present work, Nd:YVO_4 (Make: Electrox; Model: EMS 100) laser marking setup is used for the welding procedure. The two workpieces with the measurements 100 mm × 35 mm × 0.5 mm each are clipped with the assistance of a mechanical

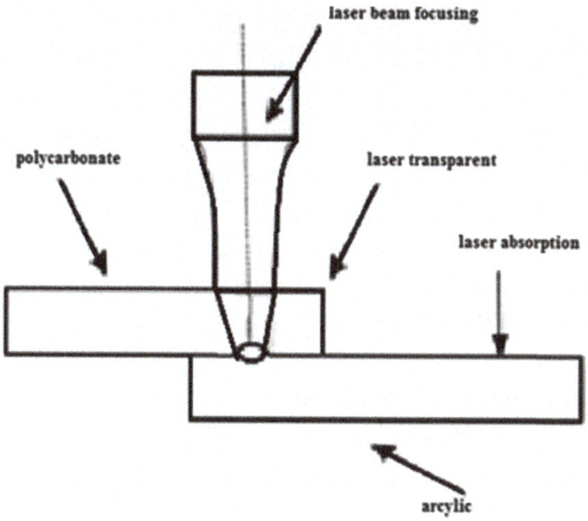

Fig. 4.1 Schematic setup of TTLW of dissimilar thermoplastics

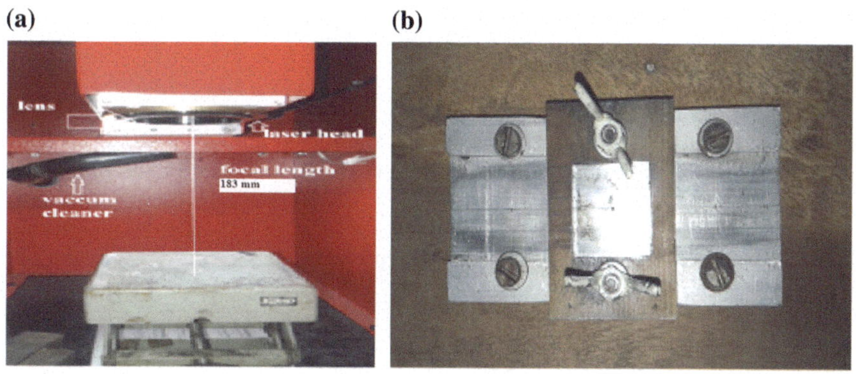

Fig. 4.2 **a** Laser welding setup and **b** mechanical setup

clamping device, as shown in Fig. 4.2b. Covering length is 70 mm. No absorbent is used between the polymer sheets. The laser used with spot diameter is around 60 μm. The polycarbonate is put at best and TTLW process is shown in Fig. 4.1. Experimental setup is shown in Fig. 4.2a. The process parameters and their cut-off points (limits) are presented in Table 4.1.

Table 4.1 Process control parameters along with their limits

Input parameters	Symbols	Units	Levels				
			(−2)	(−1)	(0)	(1)	(2)
Laser power	P	W	9.190	9.6	10.2	10.8	11.21
Laser frequency	f	kHz	265.910	300	350	400	434.090
Welding speed	S	mm/s	0.650	1	1.5	2	2.340

4.3 Results and Discussions

The results of the pull test are shown in Table 4.2. UL is found maximum for sample no. 2 and it is minimum for sample no. 7. The images of both the welded samples (sample nos. 2 and 7) and two broken samples (sample nos. 2 and 7) after pull test are shown in Fig. 4.3a and b, respectively. Load vs extension curves obtained by pull test for sample nos. 2 and 7 are shown in Fig. 4.4a and b, respectively. Using BLUEHILL software, ultimate load has been determined after the pull test.

4.3.1 Analysis-of-Variance

Table 4.3 shows the ANOVA response and the significant models ($p < 0.05$) at 95% certainty level are additionally shown. The test for significance of the regression

Table 4.2 The experimental value of ultimate load as per design

Exp. no.	P (W)	f (kHz)	S (mm/s)	UL (N)
1	9.6	400	2	280.48
2	10.8	400	1	417.33
3	10.2	434.090	1.5	354.21
4	11.20	350	1.5	344.00
5	9.6	300	2	274.96
6	9.6	400	1	352.57
7	9.190	350	1.5	215.00
8	10.2	350	1.5	282.26
9	9.6	300	1	283.02
10	10.2	265.910	1.5	268.20
11	10.2	350	2.340	337.26
12	10.2	350	0.650	383.92
13	10.8	300	1	296.31
14	10.8	400	2	330.16
15	10.8	300	2	332.26

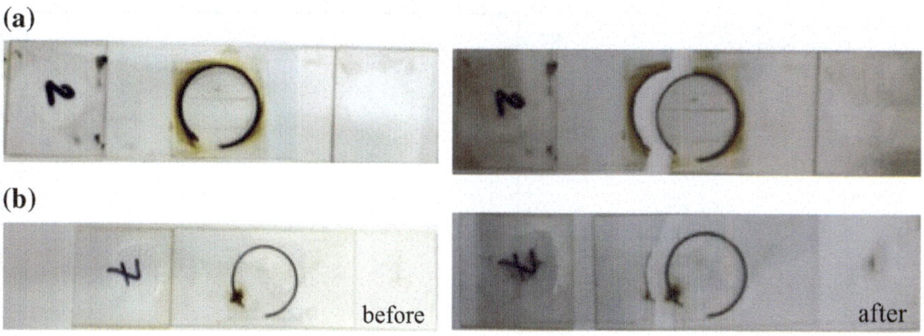

Fig. 4.3 Photographic view of the welded sample before and after the pull test. **a** Maximum UL for experiment no. 2 and **b** minimum UL for experiment no. 7

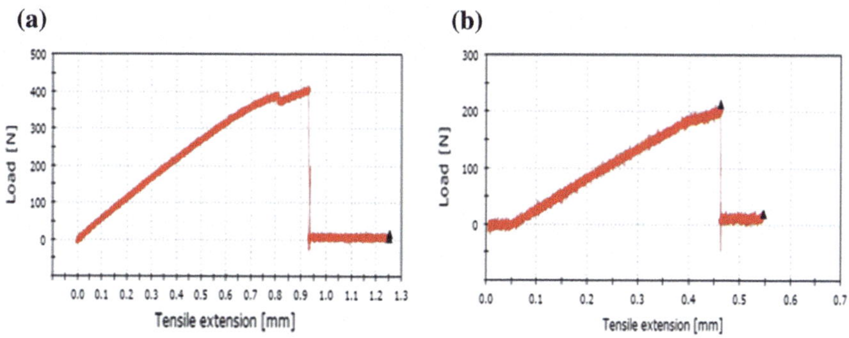

Fig. 4.4 Load versus extension curves **a** for sample no. 2 and **b** sample no. 7

Table 4.3 Analysis-of-variance (ANOVA) of ultimate load

Source	Sum of squares	df	Mean square	F-value	*p*-value	Remarks
Model	36174.61	9	4019.40	16.20	0.0034	Significant
P	11832.08	1	11832.08	47.69	0.0010	Significant
f	8397.08	1	8397.08	33.84	0.0021	Significant
S	3224.30	1	3224.30	13.00	0.0155	Significant
Pf	240.35	1	240.35	0.9687	0.3702	Non-significant
PS	104.62	1	104.62	0.4216	0.5448	Non-significant
fS	4378.14	1	4378.14	17.65	0.0085	Significant
P^2	1.71	1	1.71	0.0069	0.9371	Non-significant
f^2	690.17	1	690.17	2.78	0.1562	Non-significant
S^2	4792.47	1	4792.47	19.32	0.0071	Significant
Residual	1240.58	5	248.12			
Total	37415.20	14				

$R^2 = 0.9668$, Adjusted $R^2 = 0.9072$, Predicted $R^2 = 0.7085$, Adeq Precision = 14.3179

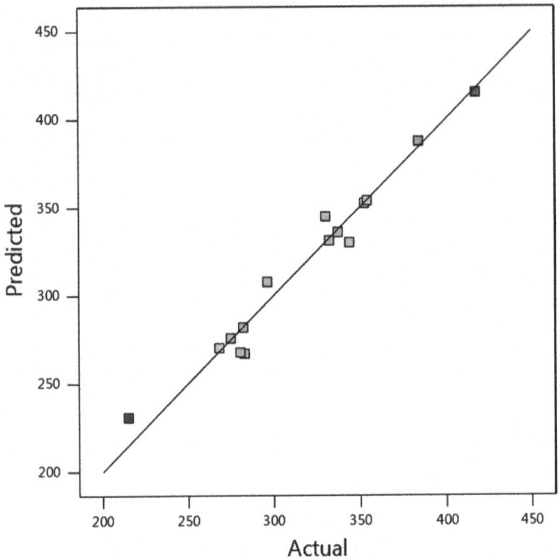

Fig. 4.5 Actual versus predicted plot for UL

model, significance test on individual model coefficients and statistical software Design-Expert V11 have been used to conduct the lack-of-fit test.

A quadratic model is chosen to correlate the input parameters with the response (Eq. 1). The F statistics and p values are used to evaluate the major factors. It can be seen in addition from the ANOVA tables that the other adequacy measure R2, adequacy precision (R2), predicted value (R2) also balanced R2 for the created model. The whole adequacy measures are near 1, which show adequate model. The satisfactory accuracy is compared and the scope of the predicted value at the design points to the average predicted error [8]. In the present work, the value of acceptable precision is considerably greater than 4. This shows the acceptability of the model. The analysis performed by ANOVA shows that for the UL model, the laser frequency f, welding speed S, laser power P, the correlation between laser frequency and welding speed $f \times S$, the quadratic effect of the welding speed S^2 are the major model terms. The rest of the model parameters are not that much significant.

$$\begin{aligned} \text{UL} =& 622.46 - 2.87\text{P} - 2.95\text{f} - 163.83\text{S} + 0.18\text{Pf} + 12.05\text{PS} - 0.93\text{fS} - 1.47\text{P}^2 \\ & + 0.004\text{f}^2 + 112.55\text{S}^2 \end{aligned} \tag{1}$$

Figure 4.5 shows the connection between the actual and predicted values of response. This figure plainly demonstrates that the created model is satisfactory and the predicted results are in great concurrence with actual data.

4.3.2 *Impact of Input Welding Parameters on Ultimate Load*

A three-dimensional response surface is constructed to find out the impact of welding input parameters on responses that have been plotted according to the fitted quadratic model. Figure 4.6a–c demonstrates the plots, with one variable kept constant at its respective centre value and the other two are plotted within the working range. It is observed from Fig. 4.6a that with an increase in laser power and laser frequency, UL is continuously increasing. When the frequency and power are too low, the welded molten pool does not get enough heat to form a decent quality joining; consequently, less ultimate load has been obtained. From Fig. 4.6b, c, it is found that UL decreases with welding speed up to a threshold value, and after that it starts increasing.

Numerical Optimization

The enhancement part in measurable programming Design-Expert V11 searches for a combination of factor levels that fulfil the necessities put on the reaction and factors at the same time. This advancement method actualizes the structure space using

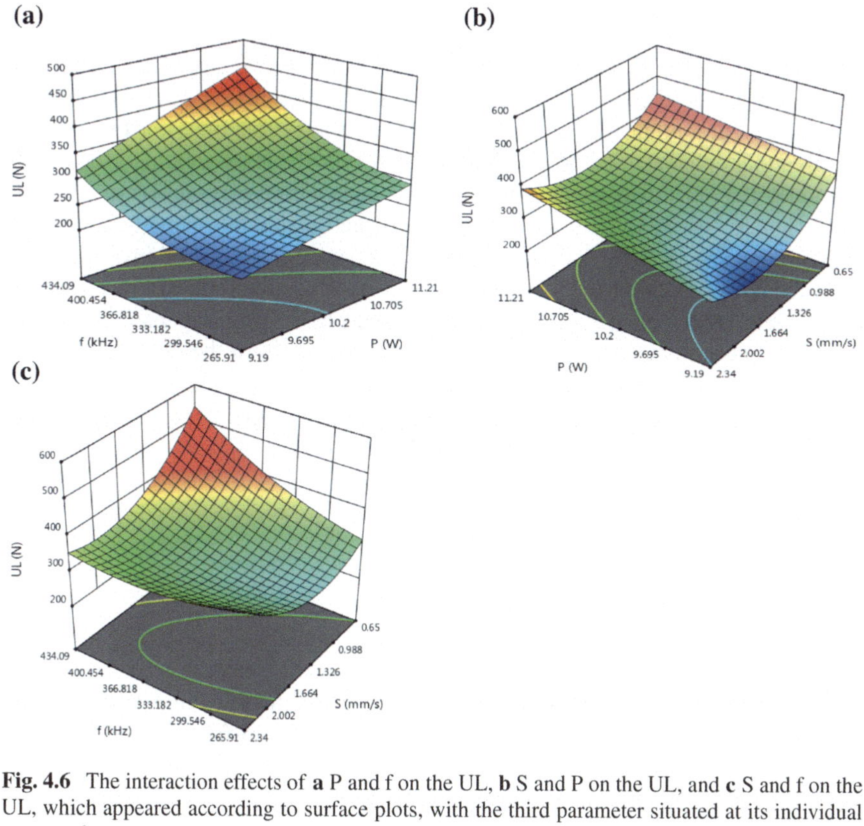

Fig. 4.6 The interaction effects of **a** P and f on the UL, **b** S and P on the UL, and **c** S and f on the UL, which appeared according to surface plots, with the third parameter situated at its individual centre value

Table 4.4 The selected criteria for numerical optimization of UL

Parameter or response	Limit		Goals	Importance
	Lower	Upper	Criterion	
P (W)	9.19	11.21	Within range	3
S (mm/s)	0.65	2.34	Within range	3
f (kHz)	265.91	437.09	Within range	3
UL (N)	215	417.33	Maximize	3

Table 4.5 Optimal welding conditions based on the selected criteria for UL

S. no.	P (W)	S (mm/s)	f (kHz)	UL (N)	Desirability
1	11.17	0.67	427.06	557.06	1.000 selected
2	11.21	0.89	425.37	501.10	1.000
3	10.38	0.78	426.61	485.129	1.000
4	10.49	0.98	425.63	442.005	1.000
5	11.03	1.01	429.26	473.566	1.000

the created model to discover the factor settings to improve any blend of at least one objective. These goals are joined into a desirability function, D. This enhancement merges at a point which amplifies this desirability function. For this current investigation, enhancement criteria have been set for the laser welding process. In the model, the objectives are changed in accordance with the expansion of ultimate load; whereas, the procedure parameters are picked inside the scope of examined configuration space. Table 4.4 outlines the criteria, upper and lower limits, speaking to the objectives and also the significance of every objective for every reaction and factor. Table 4.5 demonstrates the after-effects of the best five arrangements of pareto ideal parameter mixed for chosen criteria. It is obvious from the outcomes that to accomplish most extreme UL for the welded test, the ideal parametric range for the laser power must be 10.38–11.21 W, scanning speeds must be 0.67–1.01 mm/s and frequency inside the scope to be 425.37–429.26 kHz.

4.4 Confirmatory Test

The optimization results got by desirability function analysis are additionally validated by leading corroborative tests. Decision on welding conditions is made haphazardly from Table 4.5. The test consequences of the investigations performed at optimum settings are presented in Table 4.6. It is observed from Table 4.6 that a minor error rate is presented among optimum and experimental values. Hence, the applied optimization technique is again validated.

Table 4.6 Test results of optimization validation

Optimum condition				
P (W)	S (mm/s)	f (kHz)		UL (N)
11.17	0.67	427.06	Avg. actual Predicted $\|Error\ \%\|$	514.03 557.06 8.37

4.5 Conclusions

From the preceding analysis and discussion, the following conclusions have been drawn:

- Laser power is the most significant factor affecting the ultimate load followed by frequency and welding speed.
- The optimized process parameters (scanning speed, laser power and frequency) for ultimate load have been evaluated from desirability analysis, and the result of the obtained parametric combination (laser power at 11.17 W, scanning speed at 0.67 mm/s and frequency at 427.06 kHz) has been verified by confirmatory experiment.
- Ultimate load increases with increasing value of laser power and frequency.

References

1. Coelho, J.P., Abreu, M.A., Pires, M.C.: High-speed laser welding of plastic films. Opt. Lasers Eng. **34**, 385–395 (2000)
2. Haberstorh, E., Hoffman, W.M., Poprawe, R., Sari, F.: Laser transmission joining in microtechnology. Microsyst. Technol. **12**, 632–639 (2006)
3. Bray, R.G., Kagan, V., Chambers, A.: Forward to better understanding of optical characterization and development of colored Polyamides for the infrared/laser welding: Part 1-efficiency of polyamides for infra-red welding. J. Reinf. Plast. Compos. **22**(6) (2003)
4. Jansson, A., Kouvo, S., Kujanpaal, V.: Quasi-simultaneous laser welding of polymers—the process and applications for mass production. In: Proceeding of ICALEO, Paper 801(2005)
5. Prabhakaran, R., Kontopoulou, M., Zak, G., Bates, P.J., Baylis, B.K.: Contour laser—laser transmission welding of glass reinforced Nylon 6. J. Reinf. Plast. Compos. **19**(2006)
6. Kumar, N., Rudrapati, R., Pal, P.K.: Multi-objective optimization in through laser transmission welding of thermoplastics using grey based Taguchi method. Procedia Mater. Sci. **5**, 2178–2187 (2014)
7. Roth, G., Rung, S., Hellmann, R.: Welding of transparent polymers using femtosecond laser. Appl. Phys. A **122**(86), 1–4 (2016)
8. Anawa, E.M., Olabi, A.G.: Using Taguchi method to optimize welding pool of dissimilar laser-welded components. Opt. Laser Technol. **40**, 379–388 (2008)

Chapter 5
Thermo-Mechanical Modeling of Friction Stir Welding of High Strength Aluminum Alloy 7075 T651

Md. Parwez Alam and Amar Nath Sinha

Abstract A finite-element-based three-dimensional thermo-mechanical model was investigated to predict the temperature distribution in tool and workpiece during friction stir welding (FSW) of AA7075 T651. ANSYS APDL had been used for this simulation. It was defined by four different load steps: plunge, dwell, traverse, and pull out of tool. Temperature-dependent friction coefficient between tool and workpiece was considered. The cooling effect caused by the free convection in ambient air, backing plate, and clamping bar has been taken into account. It was observed that the peak temperature of outside the shoulder region sharply reached its peak point and then decreased slowly, whereas the peak temperature within the shoulder region was dependent on shoulder diameter and welding speed. In addition, it is also observed that peak temperature during welding is not uniform due to temperature-dependent material properties.

Keywords Friction stir welding · Temperature · ANSYS · CONTA174 · TARGE170 · Aluminum alloy

5.1 Introduction

Manufacturing industry stimulates research for new materials to increase strength and weight ratio. Aluminum alloy is the preferable material due to its outstanding properties. However, joining of heat-treatable aluminum alloy by traditional welding is difficult [1]. Heat-treatable AA7075 T651 is a type of 7075 aluminum. To achieve T651 temper, the metal is solution heat treated, stress relieved, and then artificially aged. Friction stir welding (FSW) process is considered as the ideal joining process of these materials due to low heat input. This process was invented and patented by Wayne Thomas et al. [2]. The working principle of friction stir welding is very simple. A non-consumable rotating tool inserted into a workpiece stays for some time and then moves along the welding direction. In FSW, heat is generated due to friction

Md. P. Alam (✉) · A. N. Sinha
Mechanical Engineering Department, National Institute of Technology, Patna, India
e-mail: parwez.me13@nitp.ac.in

© Springer Nature Singapore Pte Ltd. 2020

I. Singh et al. (eds.), *Trends in Manufacturing Processes*,
Lecture Notes on Multidisciplinary Industrial Engineering,
https://doi.org/10.1007/978-981-32-9099-0_5

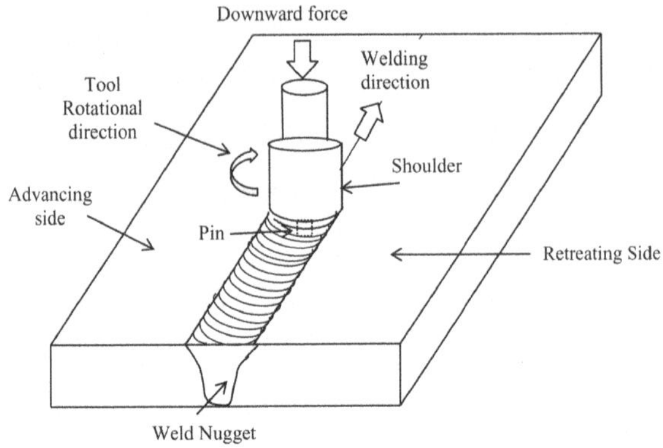

Fig. 5.1 Schematic representation of friction stir welding [7]

and plastic deformation of workpiece material. The welded joint remains with the same properties as of base metal [3]. FSW shows a fully coupled thermo-mechanical phenomena; that is plastic deformation and frictional contact are responsible for heat generation, and these temperatures are responsible for reducing the mechanical strength [4]. In earlier years, many researchers have worked on thermal analysis of friction stir welding. In 1998, Chao and Qi developed a thermo-mechanical FEM model for FSW. They considered only frictional heat as the heat source [5]. Aziz et al. developed a model by considering heat generation as a process itself by considering contact between tool and workpiece [6]. Several researchers developed model by applying external heat source in their modeling.

This paper presents thermal analysis of friction stir welding of AA7075 T651. Friction contact heat model was applied in this simulation for heat generation between tool and workpiece and different nature of temperature variation within the shoulder region and outside the shoulder region was observed. A schematic diagram of FSW is shown in Fig. 5.1.

5.2 Modeling Description and Boundary Condition

Two rectangular plates with dimension 100 mm x 50 mm x 5 mm were created and meshed. For accurate results finer mesh is created along the weld line. The total number of nodes and elements are 22,662 and 27,145, respectively. A rigid tool without pin is created on the weld joint. In this model, elements with similar thermo-physical properties of AA7075 T651 are considered. Thermal conductivity, specific heat, and density are considered as temperature-dependent, whereas Poisson's ratio and coefficient of thermal expansion are considered as temperature-independent properties in this model. It is also assumed that the workpiece behaves as rate-independent plastic material.

Selection of element in ANSYS APDL is an important parameter. In this analysis, a three-dimensional (3D) 20-node coupled-field element, SOLID226, was selected because of its large strain capabilities, large deflection, plasticity, and stress stiffness. SOLID226 is shown in Fig. 5.2. SOLID226 consists of mid-side node which can lead (a) oscillation in the thermal solution and (b) non-physical temperature distribution. Therefore, in this analysis mid-side node is not selected to avoid these phenomena.

Lagrangian model is applied for analyzing the transient effect of friction stir welding. Proper contact behavior between various parts, such as plates and pilot node, and tool is defined. CONTA174, a 3D 8-node element, was used for contact pair between the plates and TARGE170 is used for pair-based contact. Hence, it is applicable to 3D structural and coupled-field contact analysis. Figure 5.3 shows CONTA174 and TARGE170 element. Again, CONTA174 and TARGE170 are used for defining contact pair between workpiece and tool, respectively. Here, the workpiece is considered as deformable body, whereas tool is rigid. Coupled-transient thermal-structural analysis is performed because heat is generated due to frictional dissipated energy. Two real constants, FHTG and FWGT, are required for this analysis.

The rate of frictional dissipated energy is mathematically expressed as:

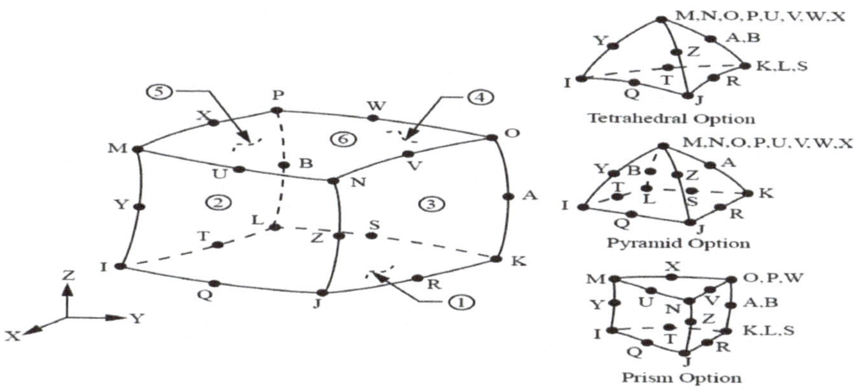

Fig. 5.2 SOLID226 [8]

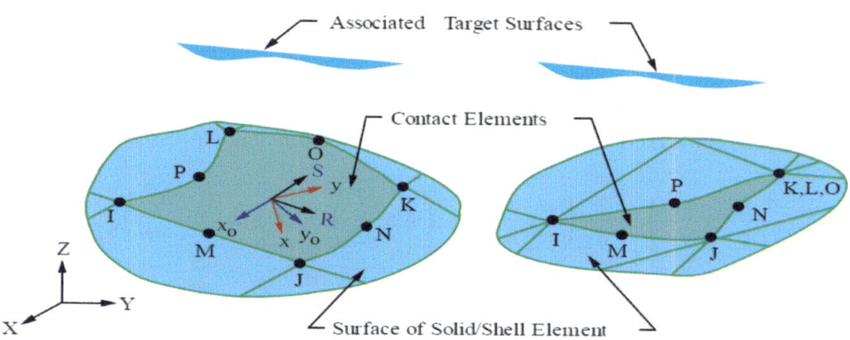

Fig. 5.3 CONTA174 and TARGE170 [8]

$$q = FHTG \times \tau \times \mathrm{V} \tag{5.1}$$

where FHTG is fraction of friction energy converted into heat. In this analysis FHTG is taken 1 for 100% frictional energy converted into heat. τ is the equivalent frictional stress and V is the sliding rate.

Heat generated during frictional contact is distributed in tool (target side, q_t) and workpiece (contact side, q_c). It is defined by real constant FWGT (heat distribution weighing factor). The target side and contact side are expressed as follows:

$$qc = FWGT \times FHTG \times \tau \times \mathrm{V} \tag{5.2}$$

$$qt = (1 - FWGT) \times FHTG \times \tau \times \mathrm{V} \tag{5.3}$$

In this analysis 5% of generated heat was distributed in tool and the remaining in workpiece.

It is considered that only 80% of plastic heat deformation converts into heat. Mathematically, it is expressed as:

$$qp = 0.8 \times \text{Plastic work} = 0.8 \int_0^t \{\sigma\}^T [\mathrm{M}]\{\mathrm{d}\varepsilon^{\mathrm{pl}}\} \tag{5.4}$$

This simulation contains four load steps: plunge, dwell, traverse, and pull out of the tool. Automatic time step is activated for faster solution. Newton Raphson technique was used to solve the governing nonlinear equation.

In thermal boundary condition, Fourier's law is applied for heat conduction in plate and heat convection is applied for heat loss to ambient and contacting surface.

The initial boundary condition for this model is used as follows:

$$\mathrm{T}(\mathrm{x},\ \mathrm{y},\ \mathrm{z}, \mathrm{t}) = \mathrm{T}_0 \tag{5.5}$$

where T_0 is the ambient temperature at time $\mathrm{t} = 0$ and T(x, y, z, t) represents the transient temperature field T, which is a function of time t and spatial coordinate (x, y, z).

In friction stir welding heat is generated by friction plastic deformation, whereas heat transfer during friction stir welding is a typical nonlinear instant heat conduction process that is governed by Fourier law of heat conduction. It is expressed as Eq. (5.6):

$$\frac{\partial}{\partial x}\left(Kx\frac{\partial T}{\partial x}\right) + \frac{\partial}{\partial y}\left(Ky\frac{\partial T}{\partial y}\right) + \left(Kz\frac{\partial T}{\partial z}\right) + \dot{Q} = \rho \mathrm{Cp}\left(\frac{\partial T}{\partial t}\right) \tag{5.6}$$

where ρ, Cp, K, and $\dot{Q}$ represent density, specific heat, directional thermal conductivity, and rate of heat generation, respectively.

In the present simulation only convective heat transfer is considered. The heat loss from free surfaces of workpiece and tool is calculated by using the Eq. (5.7)

$$q = h_{conv}(T - T_0) \tag{5.7}$$

where T, T_0, and h_{conv} refer to absolute temperature of workpiece, ambient temperature, and convection coefficient of workpiece and tool, respectively. The value of heat transfer coefficient of aluminum alloy to air is calculated as 30 W/m^2. Heat loss through backing plate and clamping bar is also considered as convection loss with high convection coefficient. In this model, heat transfer coefficient is calculated as 100 and 300 W/m^2 for clamping bar and backing plate, respectively.

The mechanical boundary condition applied as clamping bar on plate and bottom support. The bottom portion of workpiece is constraint in normal direction and taken as $U_z = 0$, whereas the clamped portion is taken as complete restraint and real, that is, there is no movement in any direction. $U_x = U_y = U_z = 0$ is applied.

5.3 Results and Discussion

In the present study, ANSYS APDL had been used for transient thermal analysis. In this process, a total of four load steps were defined: plunge stage, dwell or preheat stage, travel or welding stage, and tool plunge out stage. Heat is increased in the dwell and welding stages and the peak temperature was observed at 545.44 °C in the workpiece and 182.901 °C in the tool. In this analysis it was considered that only 5% of generated heat was distributed in tool and the remaining in workpiece [9]. Temperature distribution in workpiece and tool at 40.25 s is shown in Figs. 5.4 and 5.5, respectively.

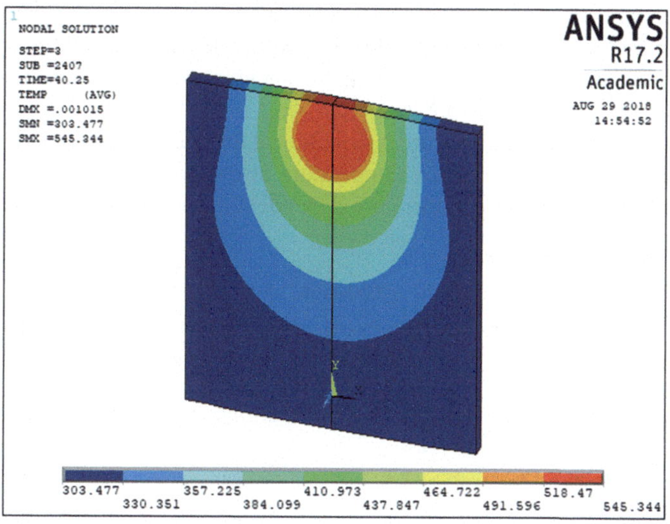

Fig. 5.4 Temperature distribution in workpiece

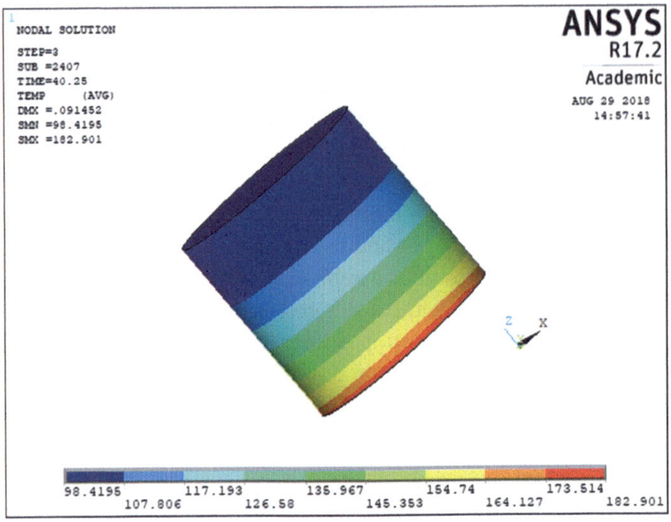

Fig. 5.5 Temperature distribution in tool

This paper describes nature and temperature distribution in workpiece at different locations. For analysis of temperature distribution, two points were chosen within the shoulder region (X_1 and X_2) and two points were chosen outside of shoulder region (X_3 and X_4). Table 5.1 shows the different locations of workpiece.

Figure 5.6 shows the temperature distribution at midpoint in Y direction (welding direction) and far from weld line (X direction). The peak temperature of X_1 and X_2 shows similar phenomena. The peak temperature at these points remains peak for some time and then gradually decreases. It is due to surface contact between tool shoulder and workpiece. The length of peak position depends on welding speed and shoulder diameter. It is also observed that this phenomenon is not happened outside the shoulder region, that is, at points X_3 and X_4.

Figure 5.7 shows the temperature distribution at the endpoint of welding. In addition, it is also observed that peak temperature at different locations of weld line was not uniform. It is due to temperature-dependent material properties and friction

Table 5.1 Temperature variation at different positions of workpiece

	X_1 (weld centre) (mm)	X_2 (border of shoulder) (mm)	X_3 (away from weld line) (mm)	X_4 (far away from weld line) (mm)
At midpoint (Y = 50 mm)	0	7.5	20	30
At endpoint (Y = 100 mm)	0	7.5	20	30

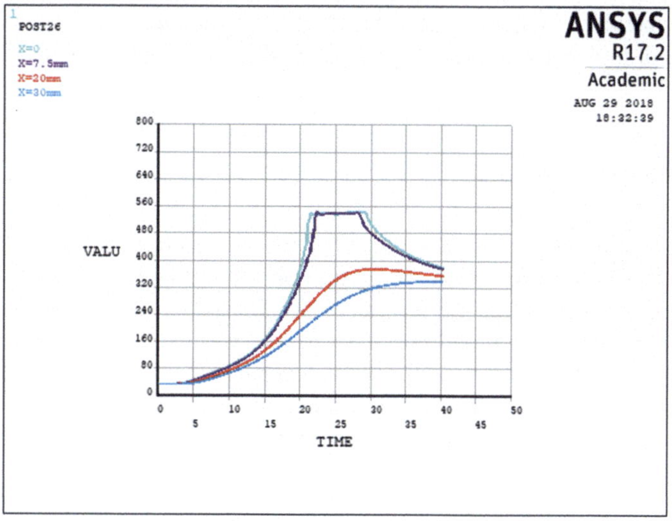

Fig. 5.6 Temperature distribution at Y = 50 mm

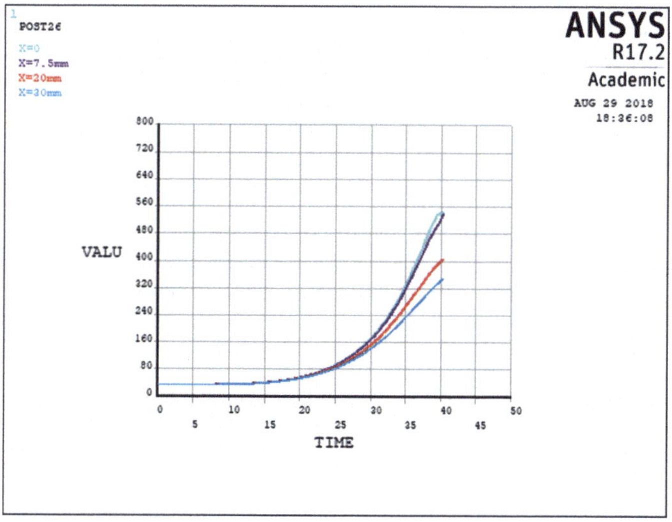

Fig. 5.7 Temperature distribution at Y = 100 mm

coefficient between tool and workpiece. As the temperature increases, the friction coefficient of tool and workpiece decreases, resulting in temperature decrease.

5.4 Conclusions

In this work, a three-dimensional model has been developed in ANSYS APDL to analyze temperature distribution during friction stir welding. It was observed that peak temperature sharply reaches its peak point and remains the same for few seconds and then decreases slowly. In addition, it is also observed that peak temperature during welding is not uniform due to temperature-dependent material properties and peak temperature within the shoulder region is almost the same. The finite-element results were similar to the findings of previous researchers.

References

1. Wahid, M.A., Khan, Z.A., Siddiquee, A.N., Shandley, R., Sharma, N.: Analysis of process parameters effects on underwater friction stir welding of aluminum alloy 6082-T6. Proc. Inst. Mech. Eng. Part B J. Eng. Manuf. 0954405418789982 (2018)
2. Thomas, W.M., Nicholas, E.D., Needam, J.C., Murch, M.G., Templesmith, P., Dawes, C.J.: GB Patent Application No. 9125978.8, December 1991 and US Patent No. 5460317 (1995)
3. Gibson, B.T., Lammlein, D.H., Prater, T.J., Longhurst, W.R., Cox, C.D., Ballun, M.C., Strauss, A.M.: Friction stir welding: process, automation, and control. J. Manuf. Process. **16**(1), 56–73 (2010)
4. Grujicic, M., He, T., Arakere, G., Yalavarthy, H.V., Yen, C.F., Cheeseman, B.A.: Fully coupled thermomechanical finite element analysis of material evolution during friction-stir welding of AA5083. Proc. Inst. Mech. Eng. Part B J. Eng. Manuf. **224**(4), 609–625 (2010)
5. Chao, Y.J., Qi, X.: Thermal and thermo-mechanical modeling of friction stir welding of aluminum alloy 6061-T6. J. Mater. Process. Manuf. Sci. **7**, 215–233 (1998)
6. Aziz, S.B., Dewan, M.W., Huggett, D.J., Wahab, M.A., Okeil, A.M., Liao, T.W.: Impact of friction stir welding (FSW) process parameters on thermal modeling and heat generation of aluminum alloy joints. Acta Metall. Sinica (Engl. Lett.) **29**(9), 869–883 (2016)
7. Alam, M.P., Sinha, A.N.: Fabrication of third generation Al–Li alloy by friction stir welding: a review. Sadhana **44**(153), 1–13 (2019)
8. Ansys Inc., ANSYS Mechanical APDL Theory Reference
9. Chao, Y.J., Qi, X., Tang, W.: Heat transfer in friction stir welding—experimental and numerical studies. J. Manuf. Sci. Eng. **125**(1), 138–145 (2003)

Chapter 6
Full-Factorial Design Approach to Parametric Optimization and Modeling of Car Suspension System

Mohd Avesh and Rajeev Srivastava

Abstract The present study is concerned with the design of an automotive suspension system that leads to better dynamic performance in uncertain operating conditions. The analysis is being carried out on a half-car suspension system model subjected to the random road excitations. Full-factorial design approach has been applied at the deliberated range of different speed levels using MINITAB® V17 tool. The major design parameters have been optimized to obtain the maximum level of comfort and vehicle body stability. A half-car structure has been modeled in Matlab Simulink to examine the dynamic behavior through the parametric variations in the ISO 2631-1:1997 recommended ranges of body acceleration. The reliable results for better dynamic performance of the proposed model have been obtained. A dynamic model of the double-wishbone suspension model generated at ADAMS® successfully validates the results.

Keywords Full-factorial · Ride comfort · Half car · Suspension system

6.1 Introduction

Uncertain road profile and shape irregularities create the chaos in vehicle body due to unnecessary vibrations transmission via the wheel and other connecting links [1]. The disturbance occurs due to road irregularities and other reasons, which is very much harmful to the vehicle as well as to the passengers [2]. Vertical body vibrations directly affect the passenger comfort that can measure in terms of the root mean square value of sprung mass acceleration [3].

In the field of vehicle dynamics, a lot of work has been done in the recent past; several approaches have been developed to explore the optimum design and parametric setting of the suspension system. A parametric optimization approach has been applied to a nonlinear car model subjected to random road excitations, in which damping and spring coefficient were the parameters of study [4]. The suspension

M. Avesh (✉) · R. Srivastava
Motilal Nehru National Institute of Technology Allahabad, Allahabad 211004, India
e-mail: mail2avesh@gmail.com

© Springer Nature Singapore Pte Ltd. 2020

I. Singh et al. (eds.), *Trends in Manufacturing Processes*,
Lecture Notes on Multidisciplinary Industrial Engineering,
https://doi.org/10.1007/978-981-32-9099-0_6

Table 6.1 Parametric search range

Factors	Search range	
	Min (−1)	Max (+1)
Sprung mass, Ms (kg)	250	300
Unsprung mass, Mu (kg)	30	60
Spring stiffness, Ks (N/m)	14000	18000
Damping coefficient, Cs (N/m-s)	500	1000

system parametric setting and optimization has been carried out by Tamboli, on the basis of real road data estimation [5]. Further, a simulation study was done on the dynamic car model subjected to real road data that was constructed at ADAMS® [6].

The present work is carried out on a half-car model to reduce the vehicle vibration through a better design of suspension system. The total five parameters, sprung mass, unsprung mass, spring stiffness, tire stiffness, and damping coefficient, have been used in analysis and design optimization. The initial search parameters value has been referred from the literature; the parameters optimization is carried out under the search range given by Table 6.1 [7]. The parametric values for the sprung mass (M_s) = 250 kg, unsprung mass (M_u) = 30 kg, spring stiffness (K_s) = 14000 N/m, tire stiffness (K_t) = 140,000 N/m, and damping coefficient (C_s) = 500 N/m-s are used in the analysis [8, 9].

6.2 Design of Experiments

Design of experiments is a significant approach to control the multiple factors together and to observe the real-time impact on the responses [10]. It facilitates the comparative design study, screening designs, and response surface modeling. The full-factorial method of experimental design is chosen for the present analysis. It is the most common and simple method of design of experiments that helps to determine the statistical significance of the factors across all possible combinations. The MINITAB®, a statistical software, is used here to implement the DOE methodology.

The experiment is conducted based on 2^5 full-factorial design with two replicates leading to a total of 64 experimental runs having several combinations of input parameters. Replication is also done to reduce variability in the model and to estimate the error. Randomization ensures each run is mutually exclusive and blocking is done to prevent any block effect and effects due to experimental factors [11]. The parametric search is undergone through the given range of parameters in Table 6.1. The search range has been decided based on various experiments conducted in the past studies and the data used by automotive manufacturers.

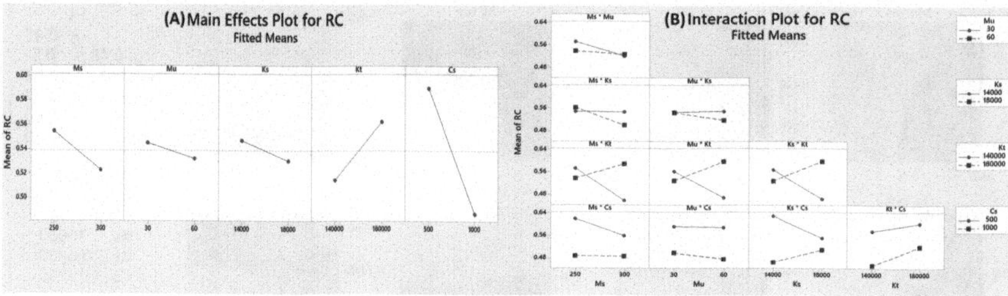

Fig. 6.1 Sensitivity analysis on ride comfort. **a** Main effect plot; **b** Interaction plot

6.2.1 Sensitivity Analysis

The sensitivity analysis is being carried out in this section to observe the change in response with the parametric variation of each factor and their interactions. This analysis has been done through the main effect plots of individual factors response shown by Fig. 6.1(a) and by interaction plots shown in Fig. 6.1(b); the line slope determines the sensitivity of the parameters on the response. The horizontal line indicates no change in response to parametric variation, and the vertical line indicates a very high parametric sensitivity. The plots show that M_u and K_s are the least sensitive parameters and K_t and C_s are identified as the highly sensitive parameters of opposite nature. The interaction plot is shown in Fig. 6.1(b) that explains the combined effect of the parameters on the response. Significant interaction means that individual parameters cannot be changed independently but changed together. Parallel lines represent no interaction, whereas non-parallel lines represent interaction. Therefore, it is spotted that the terms M_u*K_s, M_s*K_s, and K_s*K_t are highly sensitive and the terms M_s*C_s, M_u*C_s, K_s*C_s and K_t*C_s are observed the least sensitive in the region.

The two-dimensional surface plot, shown in the contour plot of Fig. 6.2, represents the relationship of three variables. The x and y axis represent the factors and the z-axis represents the response in terms of the range of color bands. The color darkness implies the elevation toward the higher value of response. The parameters C_s, M_s, and K_t find the highly sensitive parameters toward the response.

6.2.2 Response Optimizer

A response optimizer is a tool provided by MINITAB® to obtain the parametric combination for the optimal solution based on the aim of optimization. Here the goal was to maximize ride comfort, which is inversely proportional to RMS acceleration of the sprung mass. Hence the response needs to be minimized in the range suggested by ISO-2631 [12]. An upper value of 0.5 m/s^2 and the target value of 0.315 m/s^2, with weight and importance value equal to 1 were set for the model.

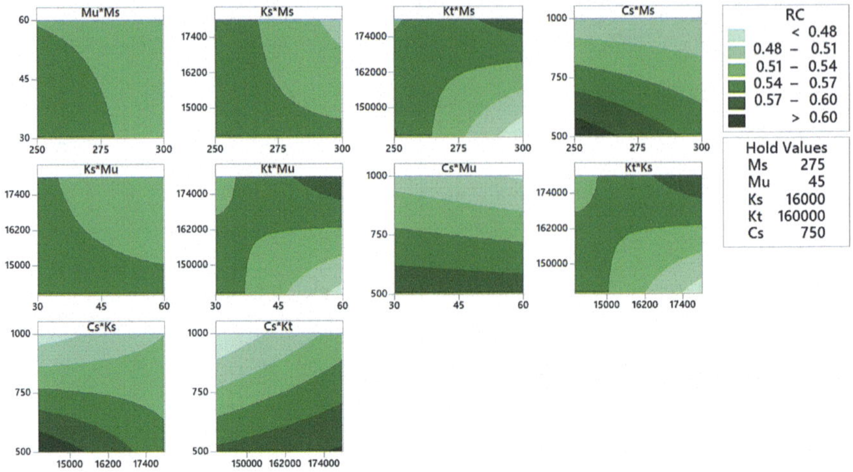

Fig. 6.2 Parametric sensitivity analysis by color bands

The optimized plot shown in Fig. 6.3 explains the effect of the individual factor on the response. In the plot, the vertical red lines show the current settings and the horizontal blue lines show the current response value. MINITAB® calculates the individual desirability (d) and composite desirability (D), which evaluates how each factor and all the factors together optimize the response. The gray region represents the area of zero desirability. The individual and composite desirability must be maximum (i.e. 1) to obtain the best results. The parameter settings can be modified by moving the cursor along the vertical bar, which recalculates the optimized factors and response (y), along with individual and composite desirability. The obtained optimized results are shown in Table 6.2.

The standard error in the estimated response is 0.0673, which is used to construct the range of 95% confidence interval. This makes sure that the actual values lie within this range with 95% confidence and it is well known that smaller the error, more precise is the result. The 95% predicted interval is used to predict the future value of the

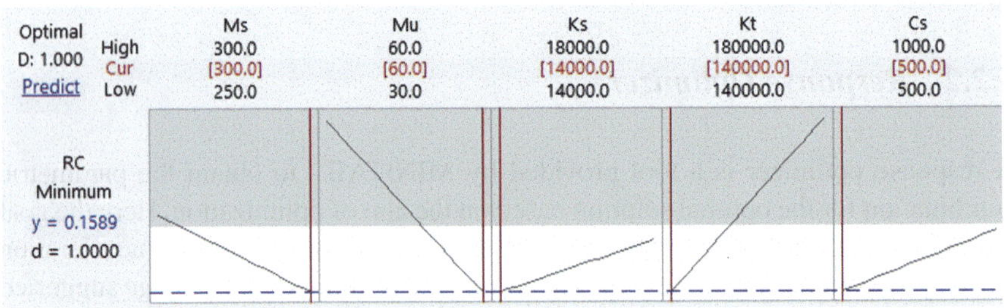

Fig. 6.3 Optimization plot

Table 6.2 Optimized results

Setting	Solution
Response = RC Goal = minimum Target = 0.315 m/s^2 Upper = 0.5 m/s^2 Weight = 1 Importance = 1	Ms = 300 kg Mu = 60 kg Ks = 1400 N/m Kt = 14000 N/m Cs = 500 N/m-s RC fit = 0.159 m/s^2 Composite desirability = 1

response for any selected combination of parameters settings. The predicted interval range is always wider compared to confidence interval range due to uncertainties involved.

6.3 Test Validation

A double-wishbone-type half-car suspension model has been constructed at ADAMS® with the optimized values of sprung mass, unsprung mass, spring stiffness, tire stiffness, and damping coefficient. The constructed model is shown in Fig. 6.4. The model excitations take place under the random road inputs, and the other parameters are taken from the literature [10, 13]. The RC was measured in terms of the RMS acceleration of sprung mass, and is obtained as 0.2150 m/s^2 with an accuracy of 70%. The obtained results have also been tested on the developed Simulink model, where the RC value obtained was 0.185 m/s^2 with an accuracy of 78%.

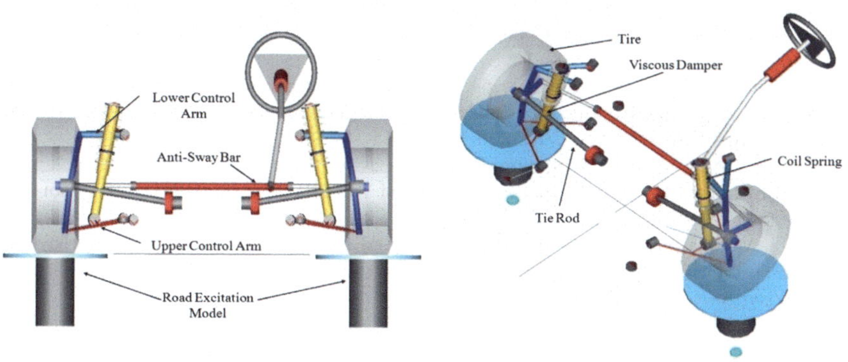

Fig. 6.4 ADAMS model

6.4 Conclusion

The ride comfort of a considered car on a road has been improved through the suspension system design optimization. The key parameters such as sprung mass, unsprung mass, spring stiffness, tire stiffness, and damping coefficient have been involved in the optimization. Design of experiments has been implemented through full-factorial methodology; the experimental model exhibited a consistent result of $R^2 = 93\%$, $R^2(\text{adj}) = 91\%$, and $R^2(\text{pred}) = 90\%$, which explains the variability, reliability, and predictability of the model.

The optimum values of input parameters have been obtained at the maximum RC that is equal to 0.1589 m/s^2. The optimized results based on response optimizer have been further validated with an accuracy level of 70% using a similar suspension model developed at ADAMS® platform. The quantified results in terms of R^2, R^2(adj), and R^2(pred) can be improved by considering various other parameters of suspension geometry, namely toe, camber, caster, tire pressure, and so on. The experimental process can be more economical by the reduction of the number of runs using fraction factorial method of DOE.

References

1. Avesh, M., Srivastava, R.: Modeling simulation and control of active suspension system in Matlab Simulink environment. In: 2012 Students Conference on Engineering and Systems, pp. 1–6. IEEE, Allahabad (2012)
2. Senthil Kumar, M., Vijayarangan, S.: Analytical and experimental studies on active suspension system of light passenger vehicle to improve ride comfort. Akt. Lengv. Automob. Pakabos Teor. Ir Eksp. Anal. Siekiant Pagerinti Važiavimo Patogumą. **65**, 34–41 (2007)
3. Koch, G., Fritsch, O., Lohmann, B.: Potential of low bandwidth active suspension control with continuously variable damper. Control Eng. Pract. **18**, 1251–1262 (2010). https://doi.org/10.1016/j.conengprac.2010.03.007
4. Verros, G., Natsiavas, S., Papadimitriou, C.: Design optimization of quarter-car models with passive and semi-active suspensions under random road excitation. Modal Anal. **11**, 581–606 (2005). https://doi.org/10.1177/1077546305052315
5. Tamboli, J.A., Joshi, S.G.: Optimum design of a passive suspension system of a vehicle subjected to actual random road excitations. J. Sound Vib. **219**, 193–205 (1999). https://doi.org/10.1006/jsvi.1998.1882
6. Gobbi, M., Mastinu, G.: Analytical description and optimization of the dynamic behaviour of passively suspended road vehicles. J. Sound Vib. **245**, 457–481 (2001). https://doi.org/10.1006/jsvi.2001.3591
7. Nagarkar, M.P., Vikhe Patil, G.J., Zaware Patil, R.N.: Optimization of nonlinear quarter car suspension–seat–driver model. J. Adv. Res. **7**, 991–1007 (2016). https://doi.org/10.1016/j.jare.2016.04.003
8. Metered, H., Kozek, M., Šika, Z.: Vibration control of active vehicle suspension using fuzzy based sliding surface. Int. J. Fuzzy Syst. Adv. Appl. **2**, 41–48 (2015). https://doi.org/10.1007/s12543-010-0056-3
9. Hemanth, K., Kumar, H., Gangadharan, K.V.: Vertical dynamic analysis of a quarter car suspension system with MR damper. J. Brazilian Soc. Mech. Sci. Eng. **39**, 41–51 (2017). https://doi.org/10.1007/s40430-015-0481-7

10. Czitrom, V.: One-factor-at-a-time versus designed experiments. Am. Stat. **53**, 126–131 (1999). https://doi.org/10.1080/00031305.1999.10474445
11. Parlak, Z., Engin, T., Çallı, İ.: Optimal design of MR damper via finite element analyses of fluid dynamic and magnetic field. Mechatronics **22**, 890–903 (2012). https://doi.org/10.1016/j.mechatronics.2012.05.007
12. Mechanical vibration and shock-evaluation of human exposure to whole-body vibration-Part 1: General requirements. Int. Organ. Stand. ISO 2631, (1997)
13. Shehata, A., Metered, H., Oraby, W.A.H.: Vibration control of active vehicle suspension system using fuzzy logic controller. In: Vibration Engineering and Technology of Machinery, Mechanisms and Machine Science. pp. 389–399 (2015)

Chapter 7
Green-Machining Characteristics Study and Comparison in Meso-Scale End-Milling of AISI-1055 and AISI-4340 Steels

Md. Zishanur Rahman, Alok Kumar Das, Somnath Chattopadhyaya and Md. Reyaz

Abstract Green-machining is an environment-friendly and cost-effective metal-cutting process because it does not use cutting fluids. This paper focusses on the study of green-machining characteristics and comparison of AISI-1055 and AISI-4340 steels in meso-scale end-milling. All the experiments were conducted with four-flute solid-carbide end-mill tool of 2 mm diameter under dry cutting condition using Taguchi L9 orthogonal array (OA) design. Machining parameters were optimized using S/N-ratio of Taguchi design technique. Analysis of variance was used for investigating the effect of cutting parameters on the burr width (BW) and surface roughness (Ra). All the experimental data obtained for "Ra" and "BW" were used to develop their regression models. The developed regression models for "Ra" and "BW" could be used for predicting the responses in meso-scale end-milling of AISI-1055 and AISI-4340 steels with minute error. At last, a comparison between AISI-1055 and AISI-4340 steels with respect to "Ra" and "BW" obtained in meso-scale end-milling was carried out. The comparative analysis between the two materials in meso-scale end-milling depicts that the machining performance of steel AISI-4340 is more effective than the machining performance of steel AISI-1055 on the basis of obtained "Ra" and "BW".

Keywords Meso-scale milling · Surface roughness · Burr width · Regression modelling

Md. Zishanur Rahman (✉) · A. K. Das · S. Chattopadhyaya
Department of Mechanical Engineering, Indian Institute of Technology (ISM), Dhanbad, India
e-mail: rzishan10@gmail.com

Md. Reyaz
Department of Mechanical Engineering, Maulana Azad College of Engineering and Technology, Neora, Patna, India

© Springer Nature Singapore Pte Ltd. 2020
I. Singh et al. (eds.), *Trends in Manufacturing Processes*,
Lecture Notes on Multidisciplinary Industrial Engineering,
https://doi.org/10.1007/978-981-32-9099-0_7

7.1 Introduction

Machining is a process of material removal from a workpiece to give a desired shape and size. The milling process is one of the conventional machining processes, which uses a rotating multi-point cutting tool for material removal. End-milling is a type of milling process which involves a mix of face milling and peripheral milling [1]. Meso-scale (1–10 mm) and micro-scale (1–1000 μm) end-milling is a scaled down process of conventional (macro) end-milling to meso/micro scale and it has been gaining a great interest in research with the rapid demands of miniaturized components in various fields of engineering and medical. It is a very useful machining method for many industries, such as biomedical, aerospace, defence, precision instruments, and so on. This process has several advantages such as low setup cost, process flexibility, ease of use, and high MRR (material removal rate), and is suitable for wide variety of engineering materials. These days, milling is one of the high added-value processes of production where accuracy and economical aspects are the most important considerations [2–5].

Green-machining relates the products/parts that are machined under healthy environment with lower costs and minimum support. On consideration of the above cases for green-machining, metal cutting process can be done either under dry cutting condition or under MQL (minimum quantity lubrication) [6–9]. Dry cutting process is ecologically more desirable because it covers the laws of health regulations and environmental protection to the maximum. For these purposes, this study has concentrated on the green-machining characteristics and comparison in meso-scale end-milling of two wear resistance and greater strength materials, AISI-1055 and AISI-4340 steels [10–14].

Surface roughness is one of the characteristics of machinability, which is commonly used to measure the machining performance. A low surface roughness of machined surface indicates the higher cutting performance and a better quality of surface generated. Machining characteristics can also be studied by analysis of burr width, burr height, over-cut, material removal rate, and so on. In this study, analysis is done based on the burr width (BW) and surface roughness (Ra) in meso-scale end-milling of steels AISI-1055 and AISI-4340 under dry cutting conditions. The relationship between machining parameters is developed with “Ra” and “BW”. Machining parameters were optimized using S/N-ratio of Taguchi technique. ANOVA was used for determining the dominant machining parameters and their effects on “Ra” and “BW”. Finally, machinability comparison between AISI-1050 and AISI-4340 steel with respect to “Ra” and “BW” was carried out.

7.2 Experimental

To conduct the meso-scale milling experiments, a three-axis CNC milling of BFW-India (Model: Chandra-plus), as shown in Fig. 7.1a, is used. The accuracy of this machine tool is 0.005 mm. For holding the end-mill cutting tool, ultra-precision

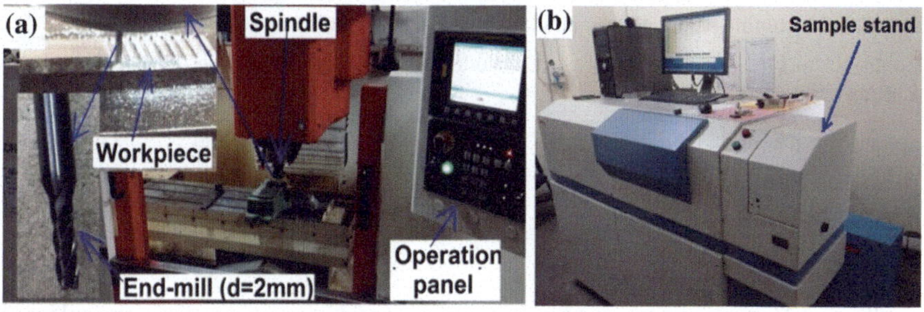

Fig. 7.1 **a** Three-axis CNC milling of BFW-India (model: Chandra-plus). **b** Spark optical emission spectrometer

Table 7.1 The chemical compositions obtained for AISI-1055 and AISI-4340 steels

Chemical composition	AISI-1055 steel		AISI-4340 steel	
	Obtained value (%)	Required range (%)	Obtained value (%)	Required range (%)
Carbon	0.512	0.50–0.60	0.430	0.35–0.45
Manganese	0.743	0.50–0.80	0.579	0.45–0.70
Silicon	0.322	0.05–0.35	0.246	0.10–0.35
Phosphorous	0.057	0.06 Max	0.032	0.05 Max
Sulphur	0.048	0.06 Max	0.032	0.05 Max
Chromium	0.110		1.082	0.90–1.40
Nickel	0.074		1.381	1.3–1.8
Molybdenum	0.019		0.215	0.20–0.35

spindle-collet is used. Four flutes with flat bottom solid-carbide end-mill (as illustrated in Fig. 7.1a) of diameter 2 mm (Model: G9A69020; K-2 carbide end-mills of YG-1 Co., Ltd.) are used for the machining, during all the experiments. All the experiments are conducted on the flat workpiece (100 × 20 × 15 mm) of both steels (AISI-1050 and AISI-4340) clamped on the table in the machine compartment of machine tool. Before the experiments, both steels are tested on spark optical emission spectrometer (a material testing machine), as shown in Fig. 7.1b. Table 7.1 presents the obtained chemical composition of both the materials, AISI-1055 and AISI-4340 steels.

7.2.1 Experimental Design

All the experiments are conducted according to Taguchi technique. It includes selection of orthogonal array, control factors and response factors (performance outputs). The selected control factors for the current investigation are three cutting parameters:

Table 7.2 Machining parameters and their values for the experiments

Control factors (cutting parameters)	Code	Levels		
		1	2	3
Spindle speed (*rpm*)	X	1600	2300	3000
Feed rate (*mm/min*)	Y	10	20	30
Depth of cut (*mm*)	Z	0.4	0.5	0.6

spindle speed in rpm, depth of cut in mm and feed rate in mm/min. Since the selected control factors are multi-level factors, three-level test for each factor was taken. The three levels for each factor (cutting parameter) are represented by 1, 2 and 3, as shown in Table 7.2. "Ra" and "BW" are chosen as response outputs for the experimentation. To accommodate three control factors and their three levels, standard Taguchi's L9 orthogonal array (OA) design was selected for achieving the objectives of how the controlled factors (cutting parameters) influence the performance outputs (Ra and BW), and what is the optimum cutting parameter to obtain low surface roughness and minimum burr width. According to the standard Taguchi's L9 (3^3) orthogonal array design (Table 7.3), nine experimental runs on meso-scale end-milling are conducted by cutting slots of 15 mm length (Fig. 7.2) on the workpiece for each material, AISI-1050 steel as well as AISI-4340 steel. In this study, the control factors (cutting parameters) are the independent variables while the response factors (Ra and BW) are the dependent variables.

Table 7.3 Results of experiments for Ra and BW

Exp. run	Machining parameters and levels			Designation	AISI-1055 steel		AISI-4340 Steel	
					Ra (1) (μm)	BW (1) (μm)	Ra (2) (μm)	BW (2) (μm)
	X	Y	Z					
1	1	1	1	X1Y1Z1	0.91	182	0.85	0
2	1	2	2	X1Y2Z2	1.59	211	1.61	0
3	1	3	3	X1Y3Z3	2.28	260	2.16	0
4	2	1	2	X2Y1Z2	0.88	165	0.70	0
5	2	2	3	X2Y2Z3	1.63	209	1.49	0
6	2	3	1	X2Y3Z1	2.05	166	1.92	0
7	3	1	3	X3Y1Z3	0.76	103	0.62	0
8	3	2	1	X3Y2Z1	0.79	115	0.68	0
9	3	3	2	X3Y3Z2	1.49	185	1.73	0

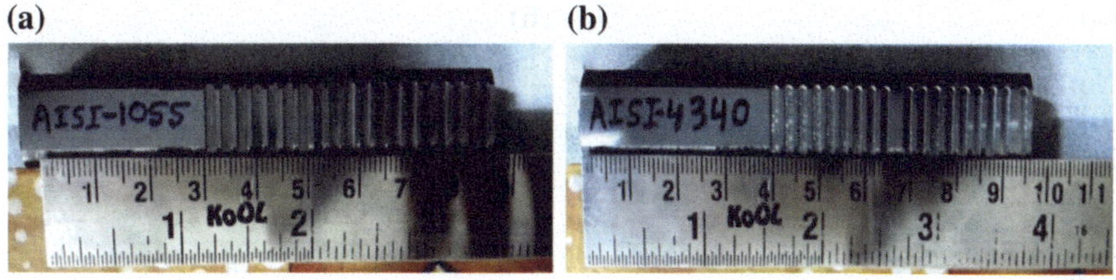

Fig. 7.2 Machined slots on the workpiece of **a** AISI-1055 and **b** AISI-4340 steel

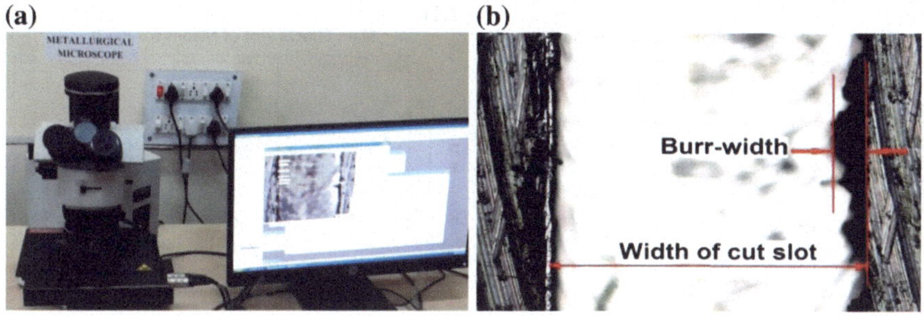

Fig. 7.3 **a** Metallurgical microscope (Model: BX51 M of OLYMPUS). **b** Burr-width sample

7.2.2 Surface Roughness and Burr Width Measurement

In this study, "Ra" and "BW" were measured for each machined slot (15 mm × 2 mm) of both the materials, as shown in Fig. 7.2a, b. A non-contact-type three-dimensional optical surface profiler of Talysurf CCI was used for measuring the surface roughness. For the measurement of burr width, a metallurgical microscope (Model: BX51 M of OLYMPUS) was used, as shown in Fig. 7.3a. Burr width of a cut slot is measured as maximum width of burr formed on one side of cut slot, as shown in Fig. 7.3b.

7.3 Results and Discussions

Under this section, S/N ratio values for each "Ra" and "BW" are calculated with the help of MINITAB-17 software. Table 7.3 describes the results of all experiments for both the materials. The optimum cutting parameters required for the minimum "Ra" and "BW" are predicted by "main effects plot of S/N ratios" drawn (Figs. 7.4 and 7.5) for "Ra" and "BW", respectively. The effect of control factors on the "Ra" and "BW" was analysed by "main effects plot of means" drawn (Figs. 7.4 and 7.5) for "Ra" and "BW" for both the materials.

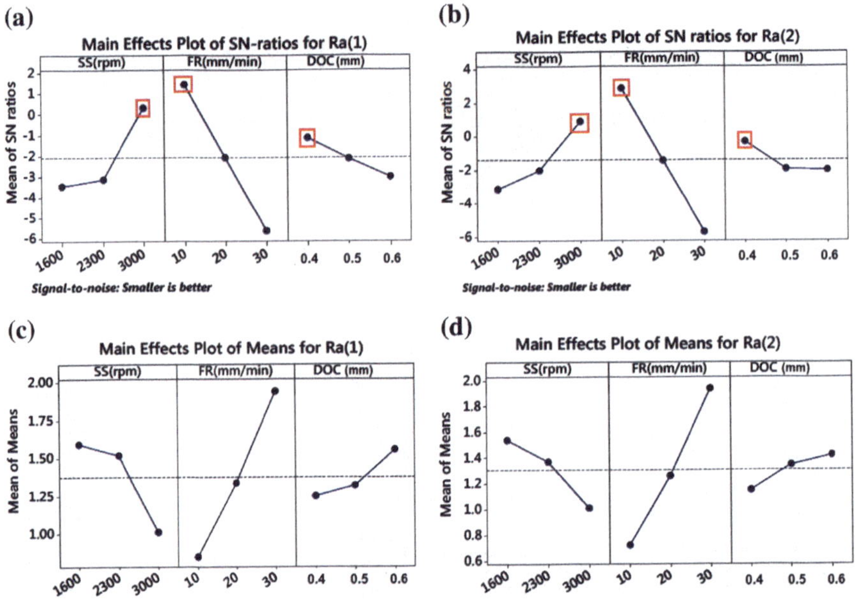

Fig. 7.4 Main effects plot of **a** S/N ratios of "Ra" for AISI-1055 steel. **b** S/N ratios of "Ra" for AISI-4340 steel. **c** Means of "Ra" for AISI-1055 steel. **d** Means of "Ra" for AISI-4340 steel

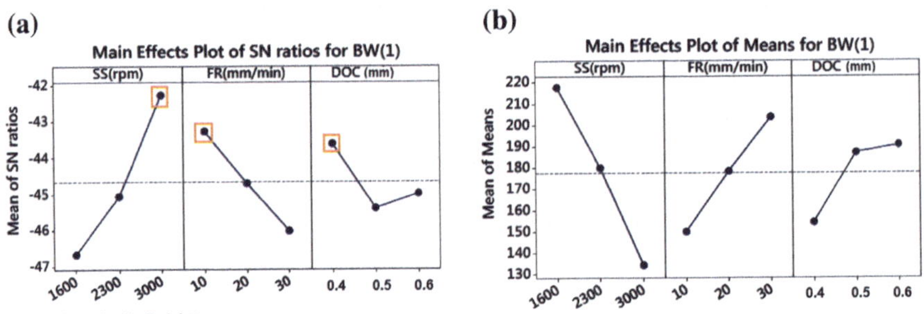

Fig. 7.5 Main effects plot of **a** S/N ratios of "BW" for steel AISI-1055. **b** Means of "BW" for steel AISI-1055

7.3.1 Average Surface Roughness

Fig. 7.4a, b reveals that the "Ra" obtained for both the steels AISI-1050 and AISI-4340 are optimum (minimum) at the spindle speed of 3000 rpm (X3), feed rate of 10 mm/min (Y1) and depth of cut 0.4 mm (Z1). From S/N ratio table, it was found that feed rate (mm/min) has more effect on "Ra", while spindle speed (rpm) has moderate effect on "Ra", and depth of cut (mm) has less effect on "Ra" in meso-scale end-milling of both the steels. From Fig. 7.4c, d, it is clear for both steels that

(i) "Ra" increases rapidly with the increase of feed rate; (ii) "Ra" decreases gradually with the increase of spindle speed, and (iii) "Ra" increases slowly when the depth of cut increases.

ANOVA is applied for significance level $\alpha = 0.05$ (i.e., confidence level = 95%). Control factors (cutting parameters) with P-value < 0.05 are acknowledged as statistically significant contribution. ANOVA results illustrate that the feed rate has more significance (69.31% for steel AISI-1055 and 77.82% for steel AISI-4340 l) on the surface roughness, which is statistically significant, while spindle speed is moderately significant (23.18% for steel AISI-1055 and 15.41% for steel AISI-4340) on the surface roughness, which is not statistically significant in meso-scale end-milling of both the steels AISI-1055 and AISI-4340. Depth of cut (DOC) has least influence on the surface roughness obtained for both the steels, which shows it is also statistically not significant.

7.3.2 Burr Width

Figure 7.5a reveals that the burr width (BW) obtained for steel AISI-1050 is minimum (optimum) at the spindle speed of 3000 rpm (X3), feed rate of 10 mm/min (Y1) and depth of cut 0.4 mm (Z1). From S/N ratio table, it was found that spindle speed (rpm) has more effect on "BW", feed (mm/min) rate has moderate effect on "BW" and depth of cut (mm) has less effect on "BW" in meso-scale end-milling of steel AISI-1055. From the main effects plot of mean for "BW" (Fig. 7.5b), it was found for AISI-1055 steel that (i) "BW" decreases rapidly with increase in spindle speed, (ii) "BW" increases gradually with increase in feed rate and (iii) "BW" increases gradually with increase in depth of cut from 0.4 to 0.5 mm and it increases very slowly with increase in depth of cut from 0.5 to 0.6 mm. It has been observed that almost negligible magnitude of burr width was formed in case of meso-scale end-milling of steel AISI-4340 l with dry cutting condition.

7.3.3 Regression Modelling

Regression modelling was done to obtain a relationship between machining parameters ("X", "Y" and "Z") and response factors ("Ra" and "BW") using the standard statistical software "MINITAB-17". After neglecting some insignificant coefficient, following are the developed regression models:

(a) Regression model of "Ra" for steel AISI-1055;

$$\begin{aligned} \mathrm{Ra}(1) &= 0.2071 + 0.000414\,\mathrm{X} + 0.07514\,\mathrm{Y} - 2.774\,\mathrm{Z} + 0.002\,\mathrm{Y} * \mathrm{Y} \\ &\quad + 23.83\,\mathrm{Z} * \mathrm{Z} + 0.000044\mathrm{X} * \mathrm{Y} - 0.004048\mathrm{X} * \mathrm{Z} - 0.4333\mathrm{Y} * \mathrm{Z} \end{aligned} \tag{7.1}$$

$$R^2 = 99.02\%$$

(b) Regression model of "Ra" for steel AISI-4340;

$$\begin{aligned} Ra(2) = 0.3019 - 0.001076\,X + 0.07733\,Y + 3.379\,Z + 0.000633\,Y * Y \\ + 2.167\,Z * Z + X * Y + 0.000190\,X * Z - 0.1900\,Y * Z \end{aligned} \quad (7.2)$$

$$R^2 = 99.12\%$$

(c) Regression model of "BW" for steel AISI-1055;

$$\begin{aligned} BW(1) = 9.238 + 0.02048\,X - 12.86\,Y + 911.7\,Z + 0.1250\,Y * Y \\ + 550\,Z * Z + 0.005714\,X * Y - 0.4\,X * Z - 8\,Y * Z \end{aligned} \quad (7.3)$$

$$R^2 = 99.21\%$$

In regression analysis, generally the R^2 value is used to validate the developed regression models and the R^2 value should lie between 0.8 and 1.0 [15]. In the present study, the developed regression models [Eqs. (7.1), (7.2) and (7.3)] are consistent because $R^2 > 90\%$. The predicted values obtained from developed regression models are compared with the experimental values for Ra(1), Ra(2) and BW(1), as depicted in Fig. 7.6a, b and c, respectively. From these figures, it is clear that the variation between experimental values and predicted values is very minimal. Therefore, the developed regression models of second-order are statistically significant for Ra(1), Ra(2) and BW(1). Hence, for further analysis, these models [Eqs. (7.1), (7.2) and (7.3)] can be utilized.

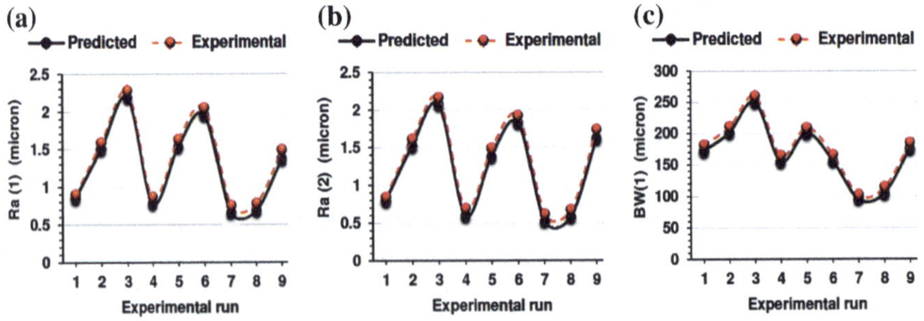

Fig. 7.6 Comparative analysis between experimental and predicted values of **a** Ra(1), **b** Ra(2) and **c** BW(1)

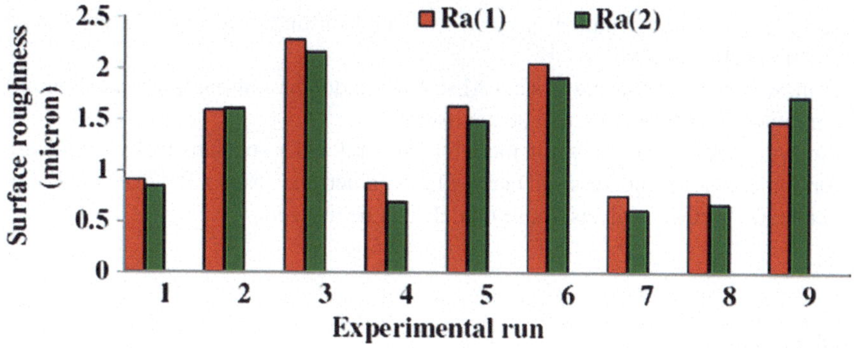

Fig. 7.7 Comparison of machinability in terms of surface roughness

7.4 Comparison Between AISI-1055 and AISI-4340 Steel

Figure 7.7 describes the comparison of machining performance in terms of "Ra" obtained in meso-scale end-milling of both the steels, AISI-1055 and AISI-4340. This figure shows that the average surface roughness [i.e., Ra(2)] found for steel AISI-4340 is lesser than the average surface roughness [i.e., Ra(1)] found for steel AISI-1055 in most of the experimental run under dry cutting condition. Moreover, almost negligible magnitude of burr width was formed in case of meso-scale end-milling of steel AISI-4340 compared to steel AISI-1055 under dry cutting.

7.5 Conclusions

This paper focussed on the comparative analysis in meso-scale end-milling of steels AISI-1055 and AISI-4340 under dry cutting condition in terms of burr width (BW) and average surface roughness (Ra) using Taguchi design method and ANOVA. This study also focussed on the regression modelling of "Ra" and "BW" obtained in meso-scale end-milling of steels AISI-4340 and AISI-1055 under dry machining. The following conclusions are summarized on the ground of experimental results and their analysis:

- Optimum solution for both surface roughness and burr width indicates that high spindle speed, low feed and low depth of cut can be used for achieving minimum surface roughness and minimum burr width in meso-scale end-milling of both steels AISI-1055 and AISI-4340 under dry cutting condition.
- ANOVA results indicate that the feed rate has more significance; spindle speed is moderate significant; and depth of cut (DOC) has least influence on the surface roughness in meso-scale end-milling of both steels AISI-1055 and AISI-4340.
- The machining performance of steel AISI-4340 is comparatively more effective than the machining performance of steel AISI-1055 on the basis of burr width and

surface roughness obtained in meso-scale end-milling under dry cutting (green machining) condition.

- In meso-scale end-milling of steel AISI-4340 under dry cutting (green machining) condition, burr formation can be controlled.
- Since the developed regression models of second-order are statistically significant for both surface roughness and burr width obtained [i.e., Ra(1), Ra(2) and BW(1)], hence, for further analysis, these models can be utilized.

References

1. Hashmi, S., Halawani, S.M., Barukab, O.M., Ahmad, A.: Model trees and sequential minimal optimization based support vector machine models for estimating minimum surface roughness value. Appl. Math. Model. **39**, 1119–1136 (2015)
2. Dhanorker, A., Özel, T.: Meso/micro scale milling for micro-manufacturing. Int. J. Mechatron. Manuf. Syst. **1**(1), 0–0 (2008)
3. Rahman, M.Z., Das, A.K., Chattopadhyaya, S., Bajpai, V., Sharma, V.: Investigation and optimization of micro-end-milling of C-103 Nb-Alloy via Taguchi design method. Am. Sci. Publ. **10**, 362–368 (2018)
4. Rahman, M.Z., Das, A.K., Chattopadhyaya, S.: Microhole drilling through electrochemical processes: a review. Mat. Man. Proc. **33**(13), 1379–1405 (2017)
5. Ozel, T., Thepsonthi, T., Ulutan, D., Kaftanoglu, B.: Experiments and finite element simulations on micro-milling of Ti–6Al–4 V Alloy with uncoated and cBN coated micro-tools. CIRP Ann. – Manuf. Technol. **60**, 85–88 (2011)
6. Ghani, J.A., Rizal, M., Hassan, C., Haron, C.: Performance of green machining: a comparative study of turning ductile cast iron FCD700. J. Cl. Prod. xxx, 1–4 (2014)
7. Heo, S.J.: Environmentally conscious hard turning of cemented carbide materials on the basis of micro-cutting in SEM: stressing four kinds of cemented carbides with PCD tools. J. Mech. Sci. Technol. **22**(7), 1383–1390 (2008)
8. Yang, S.H., Lee, Y.M., Kim, Y.S.: Analysis of cutting properties with reference to amount of coolant used in an environment conscious turning process. KSME Int. J. **18**, 2182–2189 (2004)
9. Diniz, A.E., Micaroni, R.: Cutting conditions for finish turning process aiming: the use of dry cutting. Int. J. Mach. Tools Manuf. **42**, 899–904 (2002)
10. Sreejith, P.S., Ngoi, B.K.A.: Dry machining: machining of the future. J. Mater. Process. Technol. **101**(0), x–x (2000)
11. Patel, M., Patel, J., Darji, S.: An analyse of optimum parameter on cutting force and surface roughness by taguchi method during turning on EN9 (hard steel). IJSRD **1**(3), 737–739 (2013)
12. Krishankant, J., Taneja, M., Bector, Kumar, R.: Application of Taguchi method for optimizing turning process by the effects of machining parameters. Int. J. Engg. Adv. Techn. **2**(0), 263–274 (2012)
13. Ko, J.H., Shaw, K.C., Tan, S.W., Lin, R.: Surface quality improvement in meso-scale milling with spindle axial directional ultrasonic vibration assistance. Adv. Mater. Res. **565**, 508–513 (2012)
14. Joshi, A., Kothiyal, P.: Investigating effect of machining parameters of CNC milling on surface finish by Taguchi method. IJTARME **2**(2), 2319–3182 (2013)
15. Kuram, E., Ozcelik, B.: Multi-objective optimization using Taguchi based grey relational analysis for micro-milling of Al 7075 material with ball nose end mill. Measurement **46**(6), 1849–1864 (2013)

Chapter 8
Experimental Investigation and Optimization of EDM Process Parameters in Machining of Stainless Steel s-32760 Using Taguchi Method

Renu Joshi, Satyendra Singh and Chandra Prakash

Abstract Stainless steel exposes high intensity of rigidity and strength. Electrical discharge machine (EDM) has the ability to machine such materials. Optimization of machining parameters is absolutely crucial for effective results of machining. The present work addresses the influence of process parameters, such as current (I_p), pulse duration (T_{on}), and voltage (V) on recital quantity material removal rate (MRR) on sinker electrical discharge machining of steel s-32670. Taguchi design approach has been used for planning the experiment and to establish the optimal setting level of controllable factors using signal-to-noise (S/N) ratio, and the significant contribution of these factors on output responses is analyzed with the assistance of analysis of variance (ANOVA). Minitab18 software is used to analyze the experimental design with 16 treatment condition of controllable parameters arrangement called L_{16} orthogonal array. Recital quantity MRR is calculated and amended for different set of experiments. The result of the present work reveals the optimal level of input factors for maximum MRR, while factors such as current and pulse duration play the most considerable role in EDM machining process. The findings from this study will be valuable for engineers to opt for optimal level of EDM machining factors to appliance s-32670 steel, which curtail the losses for the period of machining and directly elevate outlay and time.

Keywords Electrical discharge machine · Taguchi optimization method · Orthogonal array · Material removal rate · Steel s-32670 · Analysis of variance

8.1 Introduction

Electrical discharge machining (EDM) is the solution to incise hard materials such as steel and titanium which are difficult to machine by conventional machining. EDM's capability of using thermal energy for machining despite materials hardness has been its foremost advantage over traditional machining process in manufacturing

R. Joshi (✉) · S. Singh · C. Prakash
Department of Mechanical Engineering, B.T.K.I.T. Dwarahat, Almora, Uttarakhand, India
e-mail: joshi.renu23@gmail.com

© Springer Nature Singapore Pte Ltd. 2020

I. Singh et al. (eds.), *Trends in Manufacturing Processes*,
Lecture Notes on Multidisciplinary Industrial Engineering,
https://doi.org/10.1007/978-981-32-9099-0_8

industry. The machining arrangement of EDM requires a DC power supply, connected to shaped tool (anode) and work-piece (cathode), separated by an inter-electrode gap of 0.005–0.05 mm and located in electrically nonconducting dielectric fluid. When a voltage (200–230 V) is applied between the two electrodes and current is set at a constant value by the operator, a controlled spark is generated. Electron starts curving from anode to cathode for the pulse duration (in microseconds) of spark, which ionizes the dielectric fluids and forms a plasma channel among them. The heat generated in the plasma channel dissolves some portion of the work-piece and some of the tool material as well [1].

This machining engages thermal operations of energy generation and liberation, such as heat conduction, melting, ionization, evaporation, and collapse of gas bubbles. EDM is used to manufacture intricate cavity shapes, sharp internal corners, narrow slots, and deep ribs in molds, tools, and dies. High accuracy and precision in manufacturing of such shapes is the foremost benefit of EDM. It is playing a vital role in aerospace sector for manufacturing of high stress and temperature-gaining parts such as engine, fuel system, and gear components.

Vast research has been done on experimental and analytical studies on thermal modeling and optimization of EDM practice to progress its precision and efficiency. Chandramouli et al. [2, 3] experimentally investigated optimization of EDM process parameters in machining of 17-4 ph steel using Taguchi technique for performance factors MRR and TWR with L_{27} orthogonal array and attained optimal parameter settings for current, pulse-on time, and voltage. Reddy et al. [4] and Priyadarshini et al. [5] experimentally investigated optimization for surface roughness, MRR, and TWR of stainless steel 304 with L_9 orthogonal array and of titanium alloy with Taguchi technique and observed enhancement in all responses due to increase in current and pulse-on time. Surface roughness was measured on wire EDM for Vanadis and optimized with Taguchi L_{27} orthogonal array by Sudhakara et al. [6]. Jeykrishnan et al. [7] experimentally obtained MRR of 0.029 g/min and TWR of 0.0233 g/min for EN24 tool steel using Taguchi technique and ANOVA method with L_9 orthogonal array. Gaikwad et al. [8] analyzed optimization of MRR on Ni-Ti alloy with copper as tool material and obtained optimal MRR of 7.0806 mm^3/min. The influence of parameters and its optimization for performance factors MRR, TWR, RWR, and SR on MDN 300 steel using Taguchi method was experimented by NikelJe et al. [9] and concluded that current is more significant parameter than pulse-on time for MRR and TWR, whereas pulse-on time is more effective than discharge current for RWR and SR.

From the above literature survey, several authors have done their research on EDM process on various materials. Experimental investigation in the area of EDM of stainless steel s-32760 has been less theorized and investigated. In this paper, the influence of process parameters like pulse duration, also called pulse on time, current, and voltage to optimize the performance measure MRR on sinker electrical discharge

Fig. 8.1 Sinker EDM apparatus used for the experiment

machining EDM model 4025 for material s-32670 steel with copper alloy electrode using Taguchi optimization technique has been experimentally investigated. This approach provides a methodical algorithm to execute the experiment and optimize its machining factors. The EDM machine apparatus used for the experiment is shown in Fig. 8.1.

8.2 Schema of Investigation

Taguchi method is one of the methods used in most of the experiments for organization of parameters to get optimal performance in order to maximize enviable execution factor and minimize unenviable execution factor. It is a robust design technique that provides finest combination of process parameters in the form of orthogonal array (OA), which reduces the number of experiments as well as saves time and resources and provides superlative result for execution parameters. It also estimates the factors/parameters that most affect the product/machining quality. The investigation is done in the following sequence. The investigation steps are divided into two parts: namely (1) design of experiment, experimental setup, and procedure and (2) result and analysis.

8.2.1 *Design of Experiment, Experimental Setup, and Procedure*

8.2.1.1 Selection of Work-Piece and Electrode Material

Alloy s-32760 is used for the present set of experiment as work-piece material. It is also named as UNS s-32760, ASTM A479, and F55 according to different standards. It is a super duplex stainless steel used in aerospace and manufacturing industry, having excellent corrosion resistance, high strength with high resistance to abrasion and erosion.

The EDM operation was performed on stainless steel s-32760 of dimensions 38.4 cm length, 4 cm breadth, and 8 mm thickness. For ease in machining, the work-piece was divided into three pieces: two of them are of 12.2 cm length and one of 14 cm length. First work-piece weights 354.12 g, second work-piece weighting 368.11 g, and third 400.12 g. The dimensions of cylindrical-shaped copper electrode are: 2.2 cm diameter, length of 4 cm, and weight of 174 g. The chemical composition of work-piece material is listed in Table 8.1 and its properties are listed in Table 8.2.

Since a large amount of heat is generated in EDM while sparking, the tool should be able to bear it. Owing to high structural integrity, good conductivity of 401 W/m K, and high melting point of 1085 °C, copper electrode is used as a tool in EDM. While machining, it sears half than graphite electrode, so tool wear rate is minimum, which is required. Figure 8.2 shows the arrangement of work-piece and electrode material while machining with EDM. The tool material is properly bolted and the work-piece is clamped on magnetic plate, so that while machining both tool and work-piece should not move.

Table 8.1 Chemical composition of elements in work-piece material s-32760 steel

Alloying element	C (%)	Si (%)	Mn (%)	P (%)	Cr (%)	Ni (%)	Mo (%)
Amount	0.03	0.403	0.57	0.022	25.163	6.199	3.192
Alloying element	Cu (%)	Ti (%)	V (%)	Nb (%)	W (%)	Fe (%)	
Amount	0.52	0.006	0.087	0.025	0.667	62.946	

Table 8.2 Properties of stainless steel s-32750

Property	Quantity
Density	7.80 g/cm^3
Young's modulus	199 KN/mm^2
Specific heat	475 J/kg K
Thermal conductivity	14.2 W/m K
Tensile strength	750 N/mm^2
Hardness (Brinell)	270 max

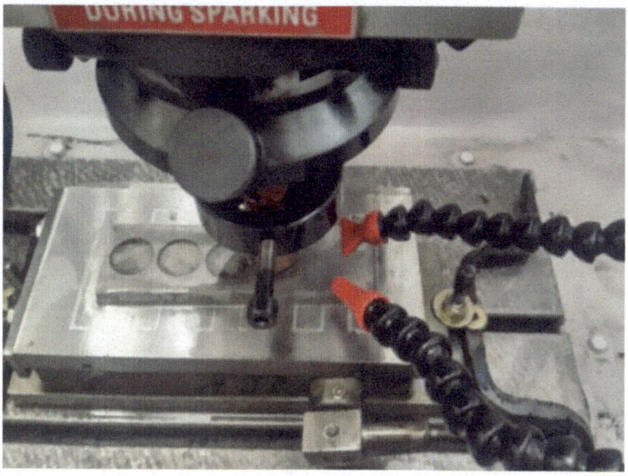

Fig. 8.2 Tool and work-piece machining arrangement

Table 8.3 Process parameters and their levels

Symbol	Process parameter	Unit	Level 1	Level 2	Level 3	Level 4
T_{on}	Pulse-on time	Microsecond	25	35	45	55
I_p	Current	Ampere	9	12	15	18
V	Voltage	Volt	40	50	60	70

8.2.1.2 Identification of Process Parameters, Determining Its Working Range and Specifying Number of Levels for Machining

The process parameters selected for the experiment are pulse duration, widely known as pulse-on time, current, and voltage, and the number of levels are 4, which specifies that four readings of each parameter are considered for the experiment. The process parameters with its levels are shown in Table 8.3. To select the range/levels of controllable parameters, trail experiments were also conducted with randomization.

8.2.1.3 Selection of Orthogonal Array

In this experiment the number of process parameter considered are 3 with four levels (shown in Table 8.3), keeping the other parameter constant. If we consider full-factorial method for design of experiment, then the design is denoted by 4^3, where 3 is the number of factors and 4 denotes how much level each factor contains. In such a case, $4 \times 4 \times 4 = 64$ experimental conditions should be obtained. To minimize these 64 set of experiment, Taguchi technique is applied. In this method, a specific set of process parameters called orthogonal array is considered for selection of design, to

Table 8.4 Experimental layout of L_{16} orthogonal array

S. no.	Current	Pulse-on time	Voltage
1	1	1	1
2	1	2	2
3	1	3	3
4	1	4	4
5	2	1	2
6	2	2	1
7	2	3	4
8	2	4	3
9	3	1	3
10	3	2	4
11	3	3	1
12	3	4	2
13	4	1	4
14	4	2	3
15	4	3	2
16	4	4	1

obtain the required number of experiments. For design of experiment of the current work, 16 set of input parameters called L_{16} orthogonal array have been considered. The experimental layout is shown in Table 8.4, where 1, 2, 3, and 4 represent the four different levels (readings) of parameters.

8.2.1.4 Conduction of Experiment Based on Orthogonal Array

Performance measure MRR needs to be calculated for the set of experiment, as presented in Table 8.4. Machining is done on work-piece with electrode using die sinker EDM model 4025 to calculate MRR. The machining surface should be fine for accurate and better machining, so before experimentation, the work-piece machining face was finished using surface grinding machine. The weight of the work-piece before and after each experiment was measured using digital electronics weight balance. Then MRR for each experiment is obtained for machining time of 20 min. The other parameters such as lift-time, duty cycle, sensitivity, sparking time, flushing pressure are kept constant.

8.2.2 *Result and Analysis*

8.2.2.1 Measure Performance/Execution Parameter MRR

The MRR is the weight loss of the work-piece under a period of machining time. More the MRR, more is the performance.

The material removal rate was calculated using the following formula:

$$MRR(mm^3/\min) = \frac{WLL(gm) \times 1000}{\rho_W(gm/cm^3) \times T(\min)}$$

where *WWL* is work-piece weight loss and calculated as:

$$WWL = (W_{\text{workpieceinitial weight}} - W_{\text{workpieceafter machining weight}})$$

ρ_w is the density of work-piece and T is the machining time in minutes.

The MRR for 16 set of experiments calculated by the above formula is listed in Table 8.5. The work-piece after machining from EDM is shown in Fig. 8.3.

The S/N ratio is represented for *higher the better* characteristics as:

$$\eta = -10\log\frac{1}{n}\sum_{i=1}^{n}\frac{1}{MRR^2}$$

Table 8.5 Experimental results of MRR and their corresponding S/N ratio

S. no.	Pulse-on time (T_{on})	Current (I_p)	Voltage (V)	MRR	S/N ratio
1	25	9	40	12.675	22.058
2	25	12	50	33.751	30.565
3	25	15	60	25.864	28.253
4	25	18	70	32.842	30.328
5	35	9	50	24.711	27.857
6	35	12	40	44.557	32.978
7	35	15	70	43.469	32.763
8	35	18	60	58.258	35.3071
9	45	9	60	23.04	27.249
10	45	12	70	35.467	30.996
11	45	15	40	40.716	32.195
12	45	18	50	64.216	36.152
13	55	9	70	40.076	32.057
14	55	12	60	36.939	31.349
15	55	15	50	57.233	35.152
16	55	18	40	64.276	36.160

Fig. 8.3 Work-piece s-32760 after EDM

where n is the number of tests and MRR is the value of MRR for *i*th test.

In the present experiment, Minitab18 software is used for applying Taguchi algorithm to analyze the set of controllable parameters and its effect on performance measure MRR. Minitab provides one-click way for S/N ratio calculation after calculating MRR experimentally. The S/N ratio for the set of experiment is listed in Table 8.5.

8.2.2.2 Find the Optimal Condition of Machining

The optimization process is divided into following steps.

(a) *Calculating S/N ratio*: The unique calculation in Taguchi method algorithm converts the quantitative assessment of recital factor (MRR in the present case) into a signal-to-noise (S/N) ratio. The S/N ratio quantifies the ability to achieve the target for a recital quantity, while minimization the variation from the target for a specific experimental condition or machine operation scenario, which means S/N ratio is used to measure the deviation of output characteristics from the desired value. The term signal represents the enviable value and noise represents the unenviable value that is measure divergence of output characteristic. The characteristic for higher value of MRR represents better machining performance called "higher the better". The characteristic that lower value represents better recital quantity such as TWR (tool wear rate) and SR (surface roughness) is called "lower the better" [10]. The S/N ratio for each set of experiment is listed in Table 8.5.

(b) *S/N Ratio Response Table and its analysis*: After calculating the S/N ratio, to know the effect of each machining parameter considering all its levels used in the experiment, response tables are provided by Minitab. It provides rank to each parameter involved in machining. The role of parameter during machining is decided by the rank. Rank 1 represents the parameter that mostly affecting the machining. The S/N response table is shown in Table 8.6.

Figure 8.4 shows MRR for different set of experiment. It is observed from the figure that at 18 A current, 55 μs pulse-on time, and 50 V voltage, the result of machining for MRR is the best. Table 8.7 shows the optimal result of considered process parameters for maximum MRR, where "A" represents pulse-on time, "B" represents current, and "C" represents voltage in level column.

(c) *ANOVA*: Analysis of variance (ANOVA) is also performed to determine the influence and contribution of machining/design parameters. Sum of squares (SS), F-test (F), degree of freedom (DF), and percentage contribution are calculated by this method, shown in Table 8.8. Minitab provides the option for ANOVA. Degree of freedom is calculated by the formula: number of levels − 1.

Table 8.6 S/N ratio response table for MRR

Level	Pulse-on time	Current	Voltage
1	27.8	27.31	30.85
2	32.23	31.47	32.43
3	31.65	32.09	30.54
4	33.68	34.49	31.54
Delta	5.88	7.18	1.89
Rank	2	1	3

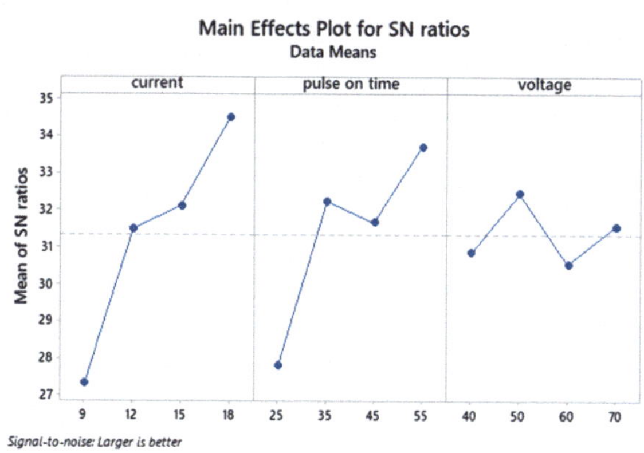

Fig. 8.4 Main effect plot for S/N ratio of MRR

Table 8.7 Optimal parameter setting of process parameters

Experimental requirement	Level	Pulse-on time	Current	Voltage
Maximum MRR	$A_4B_4C_2$	55	18	50

Table 8.8 Analysis of variance for MRR

Source	DF	Sum of squares	Contribution (%)	Mean square	F-test
Pulse-on time	3	1156.6	33.75	385.53	8.16
Current	3	1807.4	52.74	602.46	12.76
Voltage	3	179.9	5.25	59.97	1.27
Error	6	283.3	8.27	47.22	
Total	15	3427.2	100.00		

F-test is used to determine the design parameter effect on the responses. The results of ANOVA are presented in Table 8.8 showcasing current as the most significant EDM process parameter for affecting material removal rate, due to its higher percentage contribution of 52.74%, as shown in Table 8.8.

In response Table 8.6, rank 1 is given to parameter current, 2 to pulse-on time, and 3 to voltage. Similarly from ANOVA, the percentage contribution of parameters is of same order as current contribution is 52.74%, pulse-on time 33.75%, and voltage 5.25%. The response results presented in Table 8.6 and the ANOVA results presented in Table 8.8 are in good agreement.

8.2.2.3 Confirmation Test

Once the set of experiments is accomplished, an analysis is done using software by analyzing the experimental work, and an equation is developed as an outcome of experimental result called regression equation. The regression equation of the present experiment for performance measure MRR is:

$$\mathbf{MRR} = -20.2 + 0.682 \text{ Pulse} - \text{on time } + 3.115 \text{ Current } - 0.167 \text{ Voltage}$$

After obtaining the optimal level of EDM process parameters, the next step is to verify the percentage change of MRR from the experimental settings and from the regression equation, which is also called predicated MRR. Three more sets of input parameter combination of pulse-on time (T_{on}), current (I_p), and voltage (V) (which are different combination from the 16 set of experiment earlier done) are conducted and MRR is calculated. Predicted MRR from regression equation is also calculated for the three sets of experiment in order to obtain the accuracy. Percentage improvements with three sets of experiment for confirmation were recorded, as shown in Table 8.9. The average % improvement of this method is 7.53%. As a result, the prediction accuracy of this model appears satisfactory.

Table 8.9 Confirmation test results and percentage improvement

Pulse-on time (T_{on})	Current (I_p)	Voltage (V)	Experimental *MRR*	Predicted *MRR*	Error (%)
35	15	60	38.46	40.37	4.96
45	18	70	46.15	50.87	10.2
55	12	50	53.2	49.34	7.44

8.3 Conclusion

- This paper describes the application of Taguchi technique for optimization of process/input parameters for machining of s-32670 material with sinker EDM. An appropriate selection of input parameters is essential for execution of any computer-incorporated industrialized arrangement such as EDM.
- The parameters pulse duration (T_{on}) and current (I_p) have shown noteworthy effect on maximization of MRR, while parameter voltage (V) has least significant effect on machining.
- The optimal combination of input parameters for the maximum MRR of EDM is $A_4B_4C_2$, where A_4 represents pulse duration of 55 μs, B_4 represents current of 18 A, and C_2 represents voltage of 50 V.
- The result of ANOVA reveals that current has highest percentage contribution of 52.74% for best result of MRR, with pulse duration having 33.75% contribution and voltage having least contribution of 5.25%.

References

1. Joshi, S.N., Pande, S.S.: Thermo-physical modeling of die-sinking EDM process. J. Manuf. Process. **12**, 45–56 (2010)
2. Chandramouli, S., Eswaraiah, K., Optimization of EDM process parameters in machining of 17-4 PH steel using Taguchi method. Mater. Today: Proc. Part A **4**(2), 2040–2047 (2017)
3. Chandramouli, S., Eswaraiah, K., Experimental investigation of EDM process parameters in machining of 17-4 PH steel using Taguchi method. Mater. Today: Proc. **5**, 5058–5067 (2018)
4. Vikram Reddy, V., Vamshi Krishna, P., Shiva Kumar, B., Shashidhar, M.: Optimization of process parameters during EDM of stainless steel 304 using Taguchi method. Int. J. Eng. Trends Technol. **31**, 106–114 (2016)
5. Priyadarshini, M., Pal, K.: Grey-Taguchi based optimization of EDM process for titanium alloy. Mater. Today: Proc. **2**, 2472–2481 (2015)
6. Sudhakara, D., Prasanthi, G.: Application of Taguchi method for determining optimum surface roughness in wire electric discharge machining of P/M cold worked tool steel (Vanadis-4E). Procedia Eng. **97**, 1565–1576 (2014)
7. Jeykrishnan, J., VijayaRamnath, B., Akilesh, S., Pradeep Kumar, R.P.: Optimization of process parameters on EN24 tool steel using taguchi technique in electro discharge machining. Mater. Sci. Eng. **149**, 12–22 (2016)

8. Gaikwad, V., Jatti, V.S.: Optimization of material removal rate during electrical discharge machining of cryo treated NiTi alloy using Taguchi method. J. King Saud Univ. – Eng. Sci. **30**(3), 266–272 (2018)
9. NikelJe, A.M., Kumar, A., SaiSrinadh, K.V.: Influence of parameters and optimization of EDM performance measures on MDN 300 steel using Taguchi method. Int. J. Adv. Manuf. Technol. **69**(1–4), 41–49 (2013)
10. Mishra, B.P., Routara, B.C.: An experimental investigation and optimization of performance characterstics in EDM of EN.24 alloy steel using Taguchi method and grey relational analysis. Mater. Today Proc. **4**(8), 7434–7447 (2017)

Chapter 9
By-Pass Flow Meter for Sloped Pipelines

K. Kumar, K. U. Farande, S. Ajai, T. K. Sahu, A. Raut, S. Farande and Y. Tichkule

Abstract Accurate flow measurement has fiscal consequences and also necessitates fulfillment of stipulated requirement of upstream and downstream straight length at flow metering section. Many times, the site constraints do not permit to fulfill this requirement at metering sections. The present paper addresses to simulate an alternate by-pass flow meter to identify and measure the flow through the sloped pipelines of larger one for various flow rates. The alternate by-pass flow meter will be effectively useful in maintaining process accuracy by reducing downtime and improving the operational cycle of units at large. Based on the flow performance of the by-pass secondary pipes, the flow through the sloped pipe can be established by using the equation arrived by the standards flow simulation at site. The experiment reveals that by establishing by-pass arrangement, the head loss across the combined piping is minimized, The modified by-pass arrangement will be more useful for sloped pipes in terms of site constraints and operational difficulties encountered in the process industries.

Keywords By-pass flow meter · Flow measurement · Head loss · Secondary piping

9.1 Introduction

An accurate measurement of discharge through large size pipelines is necessary for effective utilization of resources. There is always a need to improve the measurement uncertainties when is charged for custody applications. Flow metering technology is typically used in various sectors, such as cooling water for thermal and atomic power

K. Kumar (✉) · K. U. Farande · S. Ajai · T. K. Sahu
Central Water & Power Research Station, Khadakwasla, Pune, India
e-mail: krishnaswamy_kumar@yahoo.co.in

A. Raut · S. Farande · Y. Tichkule
G.H. Raisoni College of Engineering, Nagpur, India

© Springer Nature Singapore Pte Ltd. 2020

I. Singh et al. (eds.), *Trends in Manufacturing Processes*,
Lecture Notes on Multidisciplinary Industrial Engineering,
https://doi.org/10.1007/978-981-32-9099-0_9

stations, lift irrigation, city water supply, inter-basin transfer of water, pump storage plants and inter-state water distribution. Measurement of discharge in hydropower plants is also important for evaluation of efficiency of the generating units, vis-à-vis for optimum water utilization.

The performance and reliability of flow meters is only achievable when uniform velocity distribution exists at flow measuring section. However, the piping configurations and fittings generate various flow distortions due to bends, valve, change in pipe cross-sections, slope, and so on. It is essential to provide minimum requirement of straight lengths of pipe upstream and downstream of flow meter as recommended by standards in order to have uniform flow at measuring section. The value of this straight length (defined in terms of diameter—D of the pipe) depends on the type of upstream and downstream disturbances.

At times, the site constraints provide inadequate upstream and downstream straight lengths, which leads to errors in flow measurement. However, the present BIS, ISO/IEC standards do not indicate the extent of deviation of measured flow rate from its true value when the recommended upstream straight lengths are not met. Thus, flow meters are susceptible to errors when the flow field at the metering section is distorted.

Mattinley and Yeh [1] studied the installation effects of various flow elements due to the upstream and downstream straight lengths. It has also been further experimented by various researchers [2–5] through their research work for the installation effects due to various elbows and bends. Hakansson and Deising [6] and Delsing et al. [7] elaborated that distorted flow will lead to increase of errors in flow metering. The effects of velocity distribution at upstream on ultrasonic flow metering accuracy were also studied [8]. Flow meter for large-scale pipes has been studied [9]. US patent application no. 2009/0320552 by Samani established the study on parallel flow meter device for measuring flow rate in pipes. The accuracy of flow measurement of flow meters is also affected by flow distortions and turbulences due to installation site conditions. Some of the typical practical site conditions are penstocks of hydropower plants, piping systems of process industries and lift irrigation systems. The piping systems are of sloped in nature. The error measured by the conventional flow measuring instruments is high and also some of the instruments may not be suitable for flow measurement due to the site conditions.

The present paper addresses to simulate an alternate by-pass flow meter to identify and measure the flow through the sloped pipelines of larger one for various flow rates. The alternate by-pass flow meter will be effectively useful in maintaining process accuracy by reducing downtime and improving the operational cycle of units at large.

9.2 Principles of By-Pass Flow Meter

The present invention uses a method that comprises a sloped main pipe and a by-pass secondary pipe. The flow is partitioned between the first pipe and the second by-pass pipe. A flow controller/control valve is connected to the by-pass flow meter which is connected to the secondary pipe. The head loss across the by-pass line is regulated by the flow controller with reference to the head loss at the sloped pipe.

It can be related as a parallel pipe. According to the principle of conservation of mass

$$Q = Q_{SM} + Q_{BP} \tag{9.1}$$

$$H_{SM} = H_{BP} \tag{9.2}$$

The subscripts SM and BP indicate the sloped main pipe and by-pass flow meter pipe. The head loss across the by-pass pipe is maintained and equated by the flow regulator as of the main sloped pipe. The head loss across the by-pass line and main pipe is measured by the high-precision digital differential pressure transmitter. A calibrated by-pass flow meter is used to measure the flow through the secondary circuit.

The total flow rate through sloped pipe and the by-pass pipe is calculated using primary flow calibration circuit. A calibration curve is established by simultaneously solving the Eqs. 9.1 and 9.2 for various flow rates through the pipelines. The established equation will help in identifying the flow through sloped pipe by measuring the flow through by-pass flow meter. The by-pass meter can be calibrated at routine interval against primary circuit without interfering in the process at large.

9.3 Experimental Investigations

9.3.1 Test Circuit

The studies were undertaken in volumetric flow meter calibration facility conforming to ISO 8316, to identify the extent of deviation in flow indicated by the by-pass flow meter, considering the head loss across the sloped pipe and the secondary circuit.

This facility at CWPRS is equipped with a primary volumetric standard with traceability to national standards and is being used for calibration of flow meters as well as for assessing performance of other flow elements, like valves and filters (Fig. 9.1). A main service pump and booster pumping units are used to pump the required flow through the test line where test flow elements are installed for calibration. The sloped piping configurations along with secondary by-pass circuit flow and head loss characteristics were simulated.

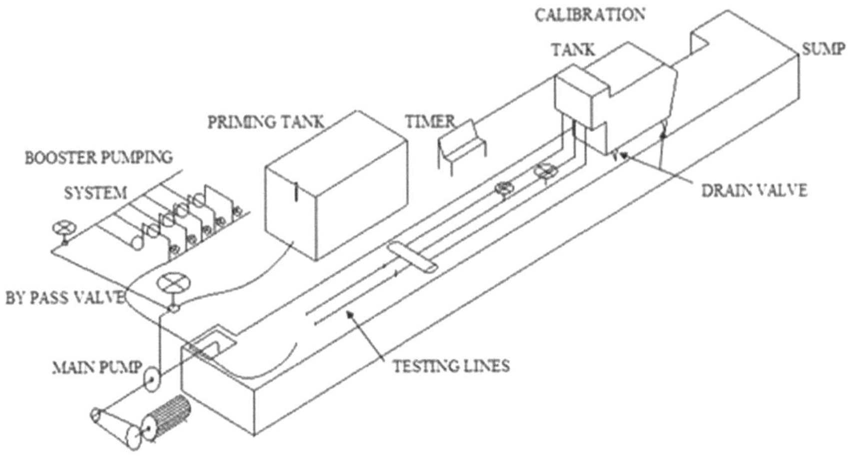

Fig. 9.1 Volumetric calibration circuit

9.3.2 Instrumentation

The instruments used for the studies are:

Flow meter: Insertion-type ultrasonic flow meter with an accuracy of 0.5%.

Diff. Pr. Transmitters: Digital transducer with an accuracy of 0.001 bar and the other parameters are:

Flow rate: ±0.3% by volumetric method

Time: 0.001 s with digital timer

Flow stability: 0.001% of set flow rate.

9.3.3 Research Studies

The sloped pipe of 100 mm NB and secondary by-pass pipe of 25 mm NB are installed in the primary flow measuring circuit, as shown in Fig. 9.2. The by-pass flow meter is connected to the secondary pipeline and the flow passes through flow controller/regulating valve. The valve maintains the constant head loss across the sloped and secondary pipelines with respect to the flow variation. The experimental studies are carried out in downstream and upstream inclination positions of the sloped pipeline (Figs. 9.3 and 9.4).

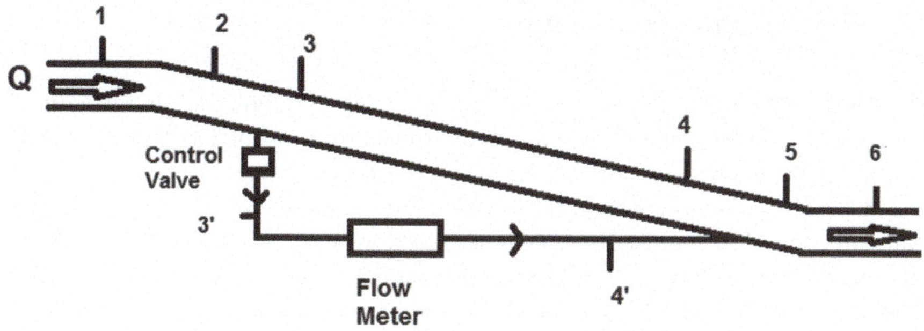

Fig. 9.2 Layout of sloped downstream large pipeline with by-pass line

Fig. 9.3 Measurement of differential pressure

Fig. 9.4 Sloped large pipeline with by-pass line

9.4 Results and Discussions

The by-pass flow meter is calibrated against primary volumetric circuit with an accuracy of ±0.5%. The by-pass flow meter is installed in secondary pipe. The head loss across the main and secondary circuits was observed using differential pressure transducers by maintaining the same pressure drop across the piping during the each run of the flow. The standard flow with by-pass flow meter fitted through the sloped pipe of downward flow is established. The standard flow tabulated (Table 9.1) in relation to the pipeline is sloped downwards, as shown in Fig. 9.5.

Table 9.1 Standard flow versus by-pass flow meter

S. no.	Collection time (s)	Head loss (bar)	Std. flow (m^3/h)	By-pass flow meter (m^3/h)	Flow through sloped pipe	% Error
1	50.957	0.0135	123.665	1.676	122.935	−0.589
2	48.009	0.0158	131.259	1.786	130.696	−0.428
3	46.652	0.0170	135.077	1.836	134.254	−0.609
4	43.284	0.0192	145.587	1.976	144.660	−0.636
5	41.933	0.0205	150.278	2.078	148.939	−0.890
6	41.090	0.0222	153.361	2.130	152.850	−0.333
7	39.216	0.0236	160.690	2.194	159.515	−0.731
8	38.669	0.0250	162.963	2.221	161.260	−1.044
9	48.734	0.0152	129.306	1.754	128.570	−0.568
10	53.008	0.0131	118.880	1.625	117.765	−0.937
11	54.858	0.0121	114.871	1.571	113.629	−1.081
12	57.577	0.0112	109.446	1.501	108.875	−0.521
13	60.172	0.0102	104.726	1.413	103.313	−1.034
14	62.866	0.0093	100.238	1.371	101.334	1.092
15	66.288	0.0082	95.064	1.281	94.802	−0.274
16	70.197	0.0075	89.770	1.195	88.869	−1.003
17	75.284	0.0066	83.704	1.114	82.790	−1.091
18	80.134	0.0053	78.683	1.039	77.979	−0.895
19	87.286	0.0046	72.195	0.945	71.549	−0.894
20	96.058	0.0038	65.602	0.850	64.891	−1.083
21	106.920	0.0030	58.937	0.758	58.298	−1.083
22	122.410	0.0025	51.479	0.644	50.978	−0.972
23	144.499	0.0018	43.610	0.546	43.157	−1.037
24	186.195	0.0011	33.844	0.416	33.497	−1.022
25	263.270	0.0006	23.935	0.333	23.692	−1.016

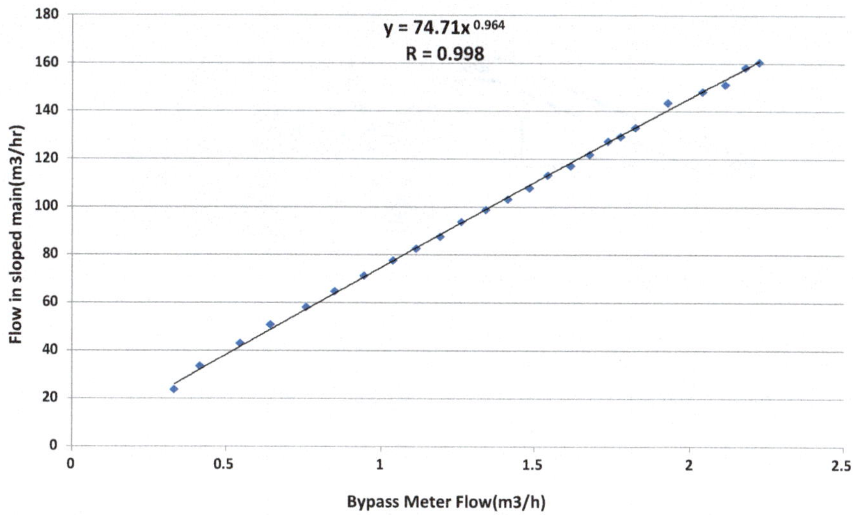

Fig. 9.5 Graphical representation of flow in sloped pipe and the by-pass meter

Table 9.2 Standard flow measured using by-pass flow meter

S. no.	Collection time (s)	Head loss (bar)	Head loss (bar) with by-pass line OpenClose		Std flow (m^3/h)	By-pass meter flow (m^3/h)	Flow through sloped pipe (m^3/h)	Total flow (m^3/h)
1	75.334	0.0061	0.0233	0.0240	83.649	1.115	82.533	83.026
2	69.679	0.0071	0.0272	0.0285	90.437	1.188	89.249	88.207
3	65.565	0.0080	0.0305	0.0321	96.112	1.264	94.848	93.640
4	62.643	0.0089	0.0341	0.0355	100.595	1.319	99.273	97.622
5	58.929	0.0100	0.0379	0.0392	106.936	1.410	105.550	104.095
6	56.983	0.0110	0.0409	0.0420	110.587	1.479	109.108	108.985
7	43.114	0.0119	0.0670	0.0708	146.161	1.984	114.177	144.655
8	42.284	0.0209	0.0745	0.0771	149.037	2.036	147.001	148.321
9	41.239	0.0215	0.0775	0.0785	152.807	2.079	150.727	151.311
10	40.036	0.0230	0.0806	0.0825	157.398	2.141	155.257	155.672
11	39.158	0.0245	0.0840	0.0865	160.928	2.184	158.744	158.642
12	38.125	0.0256	0.0875	0.0910	165.288	2.254	163.033	163.575

A graphical representation between the flow through the sloped main pipe and by-pass flow through the flow meter is established in order to arrive at the relationship between the main and secondary pipes. This graphical representation is shown in Fig. 9.5. The actual flow through the main sloped pipeline established using the relation with respect to flow through by-pass meter is given in Table 9.2.

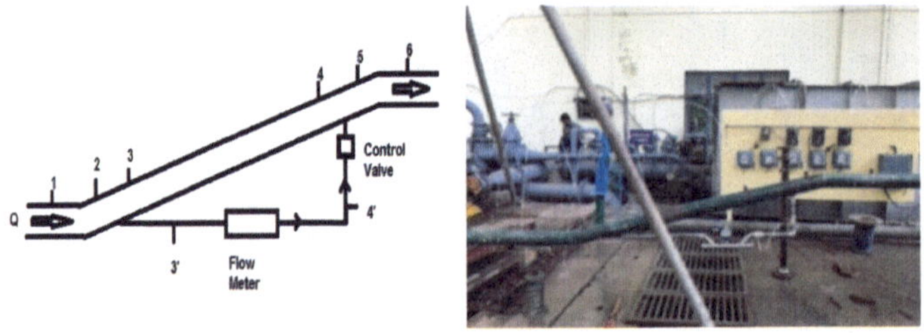

Fig. 9.6 Layout and sloped upward pipeline

Table 9.3 Standard flow versus by-pass flow meter in upward sloped pipeline

S. no.	Time (s)	Head loss (bar)	Std. flow (m^3/h)	Meter flow (m^3/h)	Flow in sloped pipe (m^3/h)	Error	% error
1	49.812	0.0115	126.502	1.316	125.186	−1.316	−1.040
2	52.519	0.0105	119.987	1.230	118.756	−1.230	−1.025
3	57.332	0.0085	109.915	1.106	108.808	−1.107	−1.007
4	60.904	0.0076	103.468	1.028	102.439	−1.028	−0.994
5	66.313	0.0065	95.028	0.937	94.091	−0.937	−0.986
6	72.868	0.0053	86.479	0.838	85.641	−0.838	−0.969
7	80.733	0.0043	78.055	0.745	77.309	−0.745	−0.954
8	91.603	0.0034	68.792	0.654	68.138	−0.654	−0.951
9	109.102	0.0024	57.758	0.538	57.220	−0.538	−0.931
10	136.787	0.0017	46.068	0.424	45.644	−0.424	−0.921
11	203.012	0.0008	31.040	0.288	30.751	−0.288	−0.930

The standard flow with by-pass flow meter fitted through the sloped pipe (Fig. 9.6) of downward flow is established. The standard flow tabulated (Table 9.3) in relation to the pipeline is sloped upwards and is shown in Fig. 9.6.

A graphical representation between the flow through the sloped inclined main pipe and by-pass flow is established in order to arrive at the relationship between the main and secondary pipe. The graphical representation is shown in Fig. 9.7. The standard flow arrived using the flow through the by-pass flow meter is given in Table 9.4.

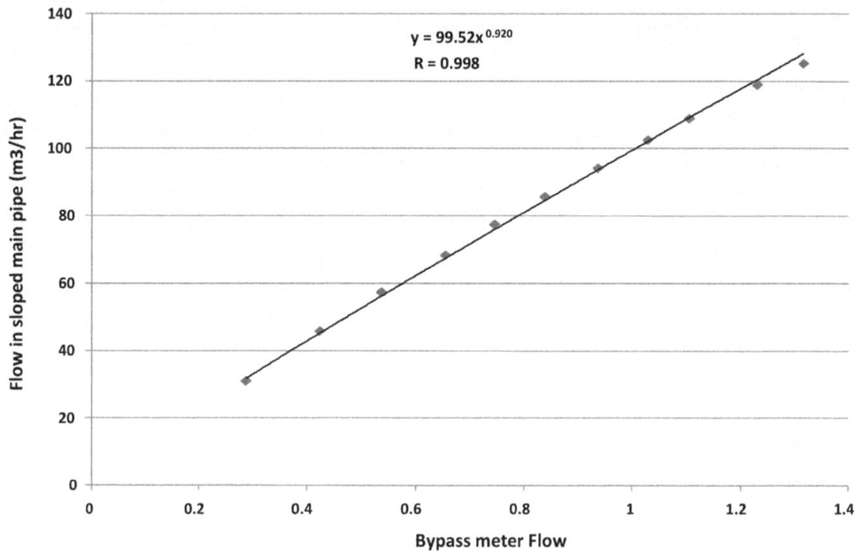

Fig. 9.7 Representation between sloped upward pipe and the by-pass line

Table 9.4 Standard flow measured using by-pass flow meter

S. no.	Collection time (s)	Head loss (bar)	Head loss with by-pass line (bar)		Std. flow (m^3/h)	By-pass flow meter (m^3/h)	Std. flow-by-pass meter flow (m^3/h)	Flow in sloped pipe (m^3/h)
			Open	Closed				
1	50.056	0.0115	0.064	0.066	125.891	1.252	124.64	122.423
2	52.244	0.0102	0.057	0.059	120.619	1.210	119.408	118.65
3	55.639	0.0091	0.051	0.052	113.259	1.116	112.142	110.121
4	60.019	0.0079	0.045	0.046	104.993	1.057	103.936	104.772
5	64.806	0.0065	0.038	0.039	97.238	0.944	96.293	94.463
6	70.904	0.0088	0.024	0.032	88.876	0.855	88.024	86.233
7	77.117	0.0045	0.027	0.028	81.715	0.768	80.946	78.075
8	86.408	0.0040	0.021	0.022	72.928	0.707	72.220	72.412
9	100.112	0.0030	0.017	0.018	62.945	0.594	62.351	61.668
10	130.202	0.0020	0.011	0.011	51.178	0.482	50.696	50.862
11	181.03	0.0009	0.004	0.005	34.800	0.292	34.508	32.107

9.5 Conclusions

Based on the flow performance of by-pass meter in secondary pipes, the respective flow through the sloped large main pipe can be established by using the equation predicted through the flow across by-pass meter. The experiment reveals that by introducing by-pass arrangement, the head loss across the combined piping is minimized for the given flow through the system. The experimental results show that the flow through the upward sloped pipeline is more than the downward sloped pipe with the same head loss measured across the pipelines. With the by-pass arrangement, the head loss is significantly reduced and this method of flow measurement will be more useful in order to overcome the site constraints, the operational difficulties encountered in the process industries and the hydropower plants.

Acknowledgements The authors are grateful to Dr. (Mrs.) V. V. Bhosekar, Director, CWPRS, Pune, for giving the guidance and motivation throughout the studies. The authors are also thankful to Shri T. Nagendra, Scientist 'E' for his continuous support and guidance during the period of investigations at various stages.

References

1. Mattingley, G.E., Yeh, T.T.: Effects of pipe elbows and tube bundles on selected types of flow meters. Flow Meas. Instrum. Butterworth-Heinemann **2**, 4–13 (1989)
2. Delsing et al.: Computer simulation approach to flow meter installation effects. In: Proceedings of FLOMEKO'93, pp. 180–187 (1993)
3. Hakansson, E., Deising, J.: Effects of flow disturbances on an ultrasonic gas flow meters. Flow Meas. Instrum. Butterworth-Heinemann **3**(4), 227–233 (1992)
4. Heritage, J.E.: The performance of transit time ultrasonic flow meters under good and disturbed flow conditions. Flow Meas. Instrum. Butterworth-Heinemann **1**, 24–30 (1989)
5. Hojholt, P.: Installation effects on single and dual beam ultrasonic flow meter. In: International Conferrence on Flow Measurement, National Engineering Laboratory, Scotland (1985)
6. Hakansson, E., Deising, J.: Effects of pulsating flow on an ultrasonic gas flow meter. Flow Meas. Instrum. Butterworth-Heinemann **5**(2), 93–101 (1994)
7. Delsing et al. Dynamic installation effects on ultrasonic flow meters. In: Proceedings of FLOMEKO'96, pp. 192–202 (1996)
8. Grimley, T.A.: Performance Testing of 12-inch Ultrasonic Flow Meters and Flow Conditioners in Short Run Meter Installations, Gas Research Institute Topical Report GRI-01/0129
9. Guo, C.-W., Wu, N.-N., Xia, S.: Flow meter for large scale pipes. In: 6th International Conference on Mining Science and Technology (2009)
10. Samani. A.Z.: Parallel flow meter device for measuring flow rate in Pipes. US Patent Publication, US 2009/0320552

Chapter 10
Natural Convection Heat Transfer from Circular Finned Helical Coil Heat Exchanger in Air

Rajesh Kumar, Prakash Chandra and Gaurav Raj

Abstract An experimental study has been conducted for natural convection heat transfer over the outer surface of circular finned and unfinned helical coil heat exchanger. Hot water flowing in the range of 1–3 l/min has been taken as working fluid and is cooled by ambient air at 32.5 ± 0.1 °C. The volume flow rate of hot water covers the Reynolds number from 4230 to 12,720 and Prandtl number around 3.7. The temperature of hot water at the inlet of test section is maintained at 50 ± 0.1 °C. The heat transfer rate of hot water has been investigated. Better outside heat transfer coefficient has been found for circular finned helical coil heat exchanger than unfinned one. It is also observed that the temperature drop decreases with increase in hot water flowing through test section.

Keywords Circular finned helical coil heat exchanger · Natural convection · Outside heat transfer coefficient · Temperature drop

Nomenclature

A	Surface area (m^2)
D	Helical coil diameter (m)
d	Tube diameter (m)
h	Heat transfer coefficient (W/m^2K)
k	Thermal conductivity (W/mK)
L	Length of tube (m)
Nu	Nusselt number (dimensionless)
P	Pitch of helical coil (m)
Pr	Prandtl number (dimensionless)
Q	Heat transfer rate (W)

R. Kumar (✉) · P. Chandra · G. Raj
Department of Mechanical Engineering, National Institute of Technology Patna, Patna 800005, India
e-mail: rajesh.me13@nitp.ac.in

© Springer Nature Singapore Pte Ltd. 2020

I. Singh et al. (eds.), *Trends in Manufacturing Processes*,
Lecture Notes on Multidisciplinary Industrial Engineering,
https://doi.org/10.1007/978-981-32-9099-0_10

Re Reynolds number (dimensionless)
T Temperature (°C)
V Volume flow rate of hot water (m^3/s)

Greek Symbols

ρ Density (kg/m^3)
C_p Specific heat (J/kg K)
θ Bulk mean temperature (°C)

Subscripts

a Ambient air
I Inlet/Inner
o Outlet/Outside
w Water

10.1 Introduction

Natural convection heat transfer from helical coil heat exchangers is extensively used in heating and cooling applications, such as domestic water heater, heat recovery system, refrigerator and air-conditioning units, cooling in electronic devices, space industry, power plant, and space vehicle. The practical application of finned helical coil heat exchanger works as an evaporator in water cooler to improve the performance. According to Zimparov [1], Dewan et al. [2], Kumbhar and Sane [3] and Liu and Sakr [4], helical coil heat exchanger is one of the most imperative passive methods that do not need external input power for enhancement of heat transfer. The heat transfer enhancement is due to the centrifugal force of fluid flow through helical coil tube and circular fins mounted on helical tube.

Ali [5] examined natural convection heat transfer from vertical helical coils of tube diameter 0.008 and 0.012 m. The results concluded that the maximum heat transfer coefficient is achieved at 0.012 m tube diameter. Ali [6] studied the steady-state free convection heat transfer from vertical helical coil tube in oil. Five diameter ratios were used in this experiment; each diameter ratio specified for two, five and ten turns. The results showed that the average heat transfer coefficient decreased with diameter ratio for number of helical coil turns of 2 or 10 and it is unstable for number of helical coil turns of 5. Prabhanjan et al. [7] experimentally performed natural convection heat transfer from helical coiled tubes in water. The outlet temperature of fluid flowing through helical tube was very close to the predicted temperature of developed model. Pawar and Sunnapwar [8] studied the steady-state convective heat transfer on helical coiled tubes in water. Correlation was developed for Prandtl number ranging from 3.81 to 4.8 and Reynolds number from 3166 to 9658 to estimate the Nusselt number.

Park et al. [9] performed natural convection heat transfer of helical coil in duct at different inclination angle. The maximum heat transfer rate of helical coil tube appeared at 80° inclination angle. Nada et al. [10] performed analysis of shell and helical coil water cooler with and without insertion of external radial fins on helical tube. The experimental results observed that the performance enhancement and compact size were achieved with the insertion of external radial fins. Andrzejczyk and Muszynski [11] experimented on shell and helical coil heat exchangers with and without external surface modification on helical tube. It was found that the shell side average Nusselt number increased with surface modification of helical coil tube. On the basis of literature review, most of the works are available on natural convection heat transfer from helical coil heat exchanger without circular fins but the present work uses circular fins on helical coil tube. The objective of the present work is to evaluate heat transfer coefficient from circular finned and unfinned helical coil heat exchanger (HCHE) experimentally for free convection.

10.2 Materials and Methods

The physical dimensions of helical coil heat exchanger are shown in Table 10.1.

Initially, both helical coil heat exchangers were constructed from straight soft copper tube of 0.00952 m outer tube diameter and eight turns with the help of wooden pattern. The tube was filled with fine sand particles before coiling to preserve the smoothness of the inner surface of tube and was flushed with compressed air. On the first helical coil heat exchanger, numbers of circular fin of 0.03 m diameter were inserted into the tube and soldered by maintaining 0.03 m distance between each fin,

Table 10.1 Physical dimensions of circular finned and unfinned HCHE

Physical dimensions	Circular finned HCHE	Unfinned HCHE
Length of tube (m)	3.8	3.8
Inner tube diameter (m)	0.009	0.009
Outer tube diameter (m)	0.00952	0.00952
Pitch (m)	0.03	0.03
Helical coil diameter (m)	0.15	0.15
Number of circular fins	121	–

(a) (b)

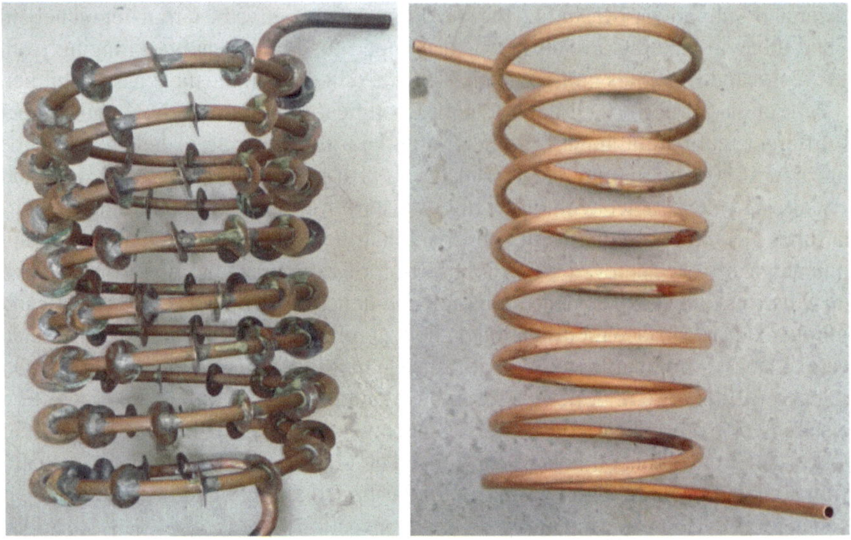

Fig. 10.1 **a** Circular finned helical coil heat exchanger. **b** Unfinned helical coil heat exchanger

as shown in Fig. 10.1a. The second helical coil heat exchanger is without circular fins, as shown in Fig. 10.1b.

10.3 Experimental Set-up and Procedure

An experimental set-up was developed to study outside heat transfer characteristics for hot water flow through helical coil heat exchanger in ambient air. Figure 10.2 shows the block diagram of experimental set-up. It consists of a 50 l galvanized iron tank, 3000 W electrical water heater is fixed at the bottom of the tank, centrifugal pump, two valves, a rotameter with an accuracy of 1.5% of full scale, four-channel digital temperature indicator and pipe for connections. In this experiment three k-type thermocouples were used; out of these, two thermocouples are inserted into 3 mm drilled hole before inlet and after outlet of each test section and remaining one was for measuring ambient temperature. A temperature controller is used to control the temperature of water in the storage tank at 50 ± 0.1 °C. Hot water flows through helical coil heat exchanger at three different flow rates, 1, 2 and 3 l/min. All temperature data were recorded at steady-state condition within ±0.1 °C by maintaining the ambient temperature at 32.5 ± 0.1 °C. This procedure is for both circular finned and unfinned helical coil heat exchanger.

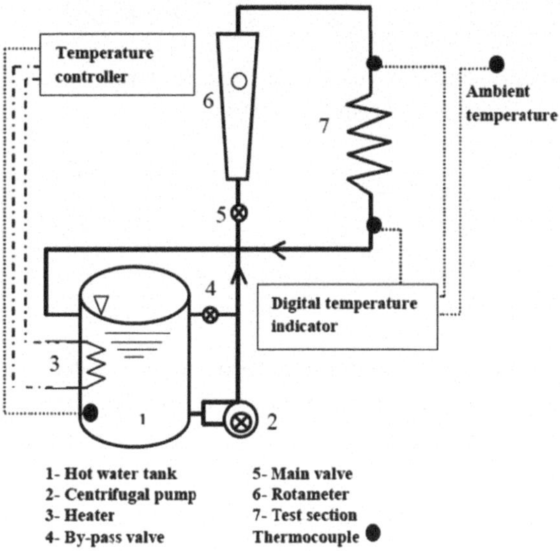

Fig. 10.2 Block diagram of experimental set-up

10.4 Analysis of Experiment

In the present work, it is assumed that the ambient air flowing on the test sections is to be constant along the length of helical coil. The inner heat transfer coefficient of circular finned and unfinned helical coil heat exchanger for turbulent flow can be calculated as Rogers and Mayhew [12] and Ali [6] from Eq. (10.1).

$$Nu = \frac{hd_i}{k} = 0.023(\mathrm{Re})^{0.85}\,\mathrm{Pr}^{0.4}\left(\frac{d_i}{D}\right)^{0.1} \tag{10.1}$$

The physical properties of water are taken from Kays et al. [13] at bulk mean temperature:

$$\theta = \frac{T_{wi} + T_{wo}}{2} \tag{10.2}$$

Total heat transfer from the helical coil heat exchanger can be calculated from Eq. (10.3):

$$Q = \rho V c_p (T_{wi} - T_{wo}) \tag{10.3}$$

where Q is the heat transfer from the helical coil, ρ is the density of water, V is the volume flow rate of water, and c_p is the specific heat of water. The properties of water were evaluated at bulk mean temperature. Once the heat transfer is obtained, the overall heat transfer coefficient is calculated from Eq. (10.4)

$$Q = UA\theta_m \tag{10.4}$$

where θ_m is the log mean temperature difference that can be calculated from Eq. (10.5) and U is defined as in Eq. (10.6)

$$\theta_m = \frac{(T_{wi} - T_{wo})}{\ln\left[\frac{(T_{wi}-T_a)}{(T_{wo}-T_a)}\right]} \tag{10.5}$$

$$\frac{1}{U} = \frac{1}{h_o} + \frac{r_o \ln\frac{r_o}{r_i}}{k} + \frac{r_o}{r_i}\frac{1}{h_i} \tag{10.6}$$

Now inner heat transfer coefficient h_i is calculated from Eq. (10.1) and outside heat transfer coefficient is obtained from Eq. (10.6).

10.5 Results and Discussion

The experiment was conducted under constant temperature (50 ± 0.1 °C) of hot water flow at the inlet of test section and is placed in ambient air at 32.5 ± 0.1 °C. The hot water flow through test sections at three different flow rates 1, 2 and 3 l/min covers the Reynolds number ranging from 4230 to 12,720 and Prandtl number around 3.7.

Figure 10.3 shows the variation of outside heat transfer coefficient with volume flow rate of hot water for circular finned as well as unfinned HCHE. As hot water flow rate increases from 1 to 3 l/min, the outside heat transfer coefficient increased for both circular finned and unfinned HCHE. A higher outside heat transfer coefficient was observed in case of circular finned HCHE as compared with unfinned HCHE.

Figure 10.4 clearly illustrates the effect of hot water volume flow rate on temperature drop for circular finned and unfinned HCHE. The temperature drop of hot

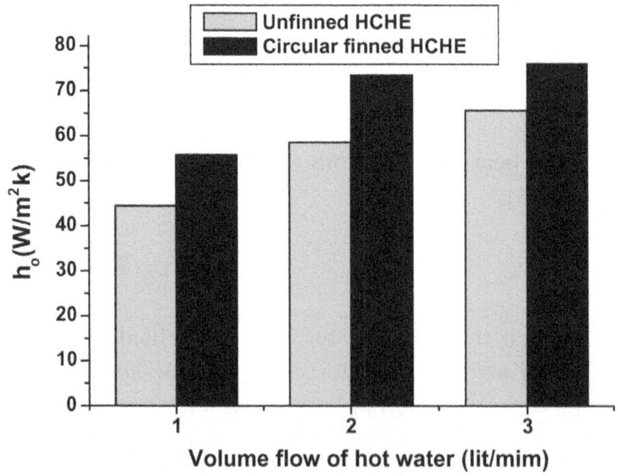

Fig. 10.3 Hot water flow through HCHE versus outside heat transfer coefficient

Fig. 10.4 Hot water flow through HCHE versus temperature drop

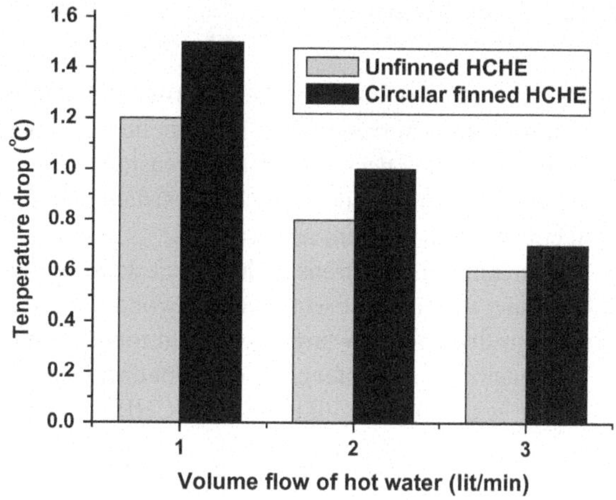

water decreases with increasing flow rate of that water with ambient temperature kept constant. The circular finned HCHE has higher temperature drop as compared with unfinned HCHE.

The bar diagram shows the effect of Reynolds number on inner heat transfer coefficient in Fig. 10.5. The present work was found to be higher inner Nusselt number as compared to Yieldiz et al. [14]. The enhancement of inner Nusselt number is due to the circular fins attached to the HCHE and also due to water flow instead of air inside HCHE.

Fig. 10.5 Reynolds number of fluid in HCHE versus inner Nusselt number

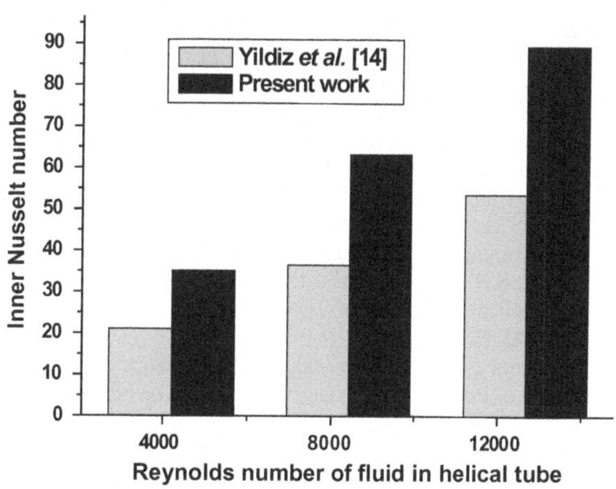

10.6 Conclusion

- The present work analyzes the experimental data for measurement of convective heat transfer of finned and unfinned helical coil heat exchanger. The temperature drop of hot water decreases with increasing volume flow rate of hot water through test section. This is due to the less resistance time available for hot water coming in contact with ambient temperature.
- The maximum heat transfer has been achieved at 3 l/min volume flow rate of hot water through test section and beyond that the temperature drop between test sections does not play any role to heat transfer enhancement.
- The outside heat transfer coefficient for designed circular finned HCHE is 15–16% higher as compared with unfinned HCHE.

References

1. Zimparov, V.: Energy conservation through heat transfer enhancement techniques. Int. J. Energy Res. **26**(7), 675–696 (2002)
2. Kumbhar, D.G., Sane, N.K.: Heat transfer enhancement in a circular tube twisted with swirl generator: a review. In: Proceedings 2010 3rd International Conference on Advances in Mechanical Engineering SVNIT Surat, pp. 188–192. Surat (2010)
3. Dewan, A., Mahanta, P., Raju, K.S., Kumar, P.S.: Review of passive heat transfer augmentation techniques. Proc. Inst. Mech. Eng. Part A J. Power Energy **218**(7), 509–527 (2004)
4. Liu, S., Sakr, M.: A comprehensive review on passive heat transfer enhancements in pipe exchangers. Renew. Sustain. Energy Rev. **19**, 64–81 (2013)
5. Ali, M.E.: Laminar natural convection from constant heat flux helical coiled tubes. Int. J. Heat Mass Transf. **41**(14), 2175–2182 (1998)
6. Ali, M.E.: Natural convection heat transfer from vertical helical coils in oil. Heat Transf. Eng. **27**(3), 79–85 (2006)
7. Prabhanjan, D.G., Rennie, T.J., Raghavan, G.V.: Natural convection heat transfer from helical coiled tubes. Int. J. Therm. Sci. **43**(4), 359–365 (2004)
8. Pawar, S.S., Sunnapwar, V.K.: Studies on convective heat transfer through helical coils. Heat Mass Transf. **49**(12), 1741–1754 (2013)
9. Park, J.H., Moon, J.Y., Chung, B.J.: Natural convection of an inclined helical coil in a duct. Int. Commun. Heat Mass Transf. **59**, 11–16 (2014)
10. Nada, S.A., Eid, E.I., Abd El Aziz, G.B., Hassan, H.A.: Performance enhancement of shell and helical coil water coolers using different geometric and fins conditions. Heat Transf. Asian Res. **45**(7), 631–647 (2016)
11. Andrzejczyk, R., Muszyński, T.: Performance analyses of helical coil heat exchangers. The effect of external coil surface modification on heat exchanger effectiveness. Arch. Thermodyn. **37**(4), 137–159 (2016)
12. Rogers, G.F.C., Mayhew, Y.R.: Heat transfer and pressure loss in helically coiled tubes with turbulent flow. Int. J. Heat Mass Transf. **7**(11), 1207–1216 (1964)
13. Kays, W., Crawford, M., Weigand, B.: Convective Heat and Mass Transfer, 4th edn. McGraw Hill, Singapore (2005)
14. Yildiz, C., Biçer, Y., Pehlivan, D.: Heat transfers and pressure drops in rotating helical pipes. Appl. Energy **50**(1), 85–94 (1995)

vhere 1 means fully disagree while 5 means fully agree. Following were the state-
ɿents:

P1. Our manufacturing system responds rapidly to changes in product volume demanded by customers
P2. Re-configures equipment to address demand changes
P3. Adjusts capacity to address demand changes
P4. Re-allocates people to address demand changes
P5. Changes manufacturing processes to address demand changes
P6. Effectively expedites emergency customer orders
P7. Responds rapidly to changes in product mix demanded by customers.

Tests were then carried out based on the data collected, as mentioned in the lowing section.

2.1 *Data Reliability Test*

very important to check the reliability of the data on a Likert scale before making analysis. This helps in understanding the internal consistency, that is, how closely ed a set of items are as a group; that is, a scale should consistently reflect the truct it is measuring.

statistics and research, internal consistency is typically a measure based on orrelations between various items of the same test (or the same subscale on a r test). It determines whether several items which propose to measure the same ruct produce similar scores.

measure is said to have a high reliability if it produces similar results under stent conditions. "It is the characteristic of a set of test scores that relates to the nt of random error from the measurement process that might be embedded in ores".

used Cronbach's alpha as a measure to check reliability of the data (refer 11.1). It is very important that all statements should be positive before perg reliability test. If there is a negative statement, the code of the statement be reversed to make it positive. In our case, all the statements were positive not require any recoding for reliability analysis. The data collected from the

.1 Cronbach's alpha

Cronbach's alpha	Internal consistency
Alpha ≥ 0.9	Excellent
$0.9 >$ alpha ≥ 0.8	Good
$0.8 >$ alpha ≥ 0.7	Acceptable
$0.7 >$ alpha ≥ 0.6	Questionable
$0.6 >$ alpha ≥ 0.5	Poor
Alpha < 0.5	Not acceptable

Chapter 11
An Investigation of Responsiveness Impact on Productivity Improvement in Indian MSME

Puneet Mangla, Ashish Agarwal and Pulak M. Pandey

Abstract The micro, small and medium enterprises (MSMEs) have a sizeable role to play in the growth of any nation. In spite of its essence in the socioeconomic progress of the nation, Indian MSMEs have to encounter big challenges in their endeavor to be competitive. Low productivity is also a challenge for the management for Indian MSMEs to sustain competition in the world market. Productivity improvement of MSMEs depends on different factors. Responsiveness of a production system is one of the major factors among them. In this paper an attempt has been made to investigate the order fulfillment strategies of Indian MSMEs with an objective to determine whether responsiveness aspects of the firms practices would affect their productivity and hence performance in terms of profits and being competitive. This investigation is based on the study of production practices adopted by North Indian MSMEs in different manufacturing sectors, like automobile components, leather products and electrical goods. In the current competitive environment enterprises must become more responsive to acquire advantages or to avoid being driven out of the market.

Keywords Responsiveness · Productivity · Performance · Competitiveness · MSME

11.1 Introduction

The MSMEs have major role to play in the social development of India and are important for the development of the country's financial system, through innovations, employment, exports, industrial output, investments, and so on. However, due to increased globalization, MSMEs have to face new challenges; moreover, many MSMEs struggle and have tough time to survive. The present producers come across

P. Mangla (✉) · A. Agarwal
Department of Mechanical Engineering, SOET, IGNOU, Maidan Garhi, Delhi 110068, India
e-mail: puneetmangla73@gmail.com

P. M. Pandey
Department of Mechanical Engineering, Indian Institute of Technology, Delhi 110016, India

© Springer Nature Singapore Pte Ltd. 2020

I. Singh et al. (eds.), *Trends in Manufacturing Processes*,
Lecture Notes on Multidisciplinary Industrial Engineering,
https://doi.org/10.1007/978-981-32-9099-0_11

a lot of challenges, few of which are, the need to rapidly adapt to the changes, variations in consumer demands, instability of vendors and materials, and increasing government regulations. Moreover, the present business environment observes severe competition because of consumer-oriented market structure and rapid up-gradation in operational technology. Thus, it is indispensable for production firms to be profitable and competitive for their survival in the market. Their performance depends on various factors, and responsiveness being one of them. In such a business setting, flourishing firms are the ones that can manage responsive and flexible production system. Thus, being responsive to the random variations in consumer demands and business environment has presented critical issues for firm's survival.

Researchers have used various approaches to study different aspects of firm's responsiveness. To fulfill the customer's requirements, companies should be able to manage the demand variations effectively and rapidly. Responsiveness of an organization refers to the capability of a system to attain its operational aims in the existence of internal and external disturbances. These disturbances are the sources of variations which do not depend on the system's intentions [1]. Responsiveness sets apart the production firms who aim at superior support to their customers with quick delivery time and properly tailored products. Moreover, major reductions in stock and cost can be attained through a better responsive operation system [2]. In today's competitive market, responsiveness has turned out to be one of the most constructive and essential capability. In the context of production, responsiveness is considered as the capability of the production firm to respond rapidly to the changes in market and customer demands. Therefore, it is vital for the production firms to accept the essential elements of responsiveness in their operations. Responsiveness in production firms has noteworthy effect on firm's competitive priorities, primarily on the delivery time. Variations in customer demands, reliable vendors, competitors and global factors (economic changes and political changes, technological developments, social changes) are the main drivers for responsiveness. Firms which are responsive results in superior customer satisfaction, reduced cost, and thus achieve higher competitive advantage [3]. Numerous production firms are following responsiveness (i.e., focused change within time due to external variations) to improve their performance and are searching for different means to improve their responsiveness. One of the means which is being adopted nowadays is the flexible production practices. Re-configurability is also used as another means to develop other systems that can augment the firm's technical capability to cater to the customers' needs by re-configuring its processes and products [4]. There is an increasing universal trend in manufacturing to utilize manufacturing practices that improve flexibility [5–11]. This practice is promoted by a premise that their utilization improves competitiveness in few performance evaluation parameters, like higher responsiveness. This is an old practice [12, 13] but its requirement is increasing as the global market is becoming highly competitive. Responsiveness can be taken as a result of, or may be inter-related to, both re-configurability [14] and flexibility [15]. Despite of the fact that these three terms do not really represent the same theory, still they are used interchangeably. In general, flexibility is understood to be an operational element, an inherent characteristic of a manufacturing system, which may be explained as

the "capability of an organization to modify status of an existing ar
the conventional parameters" [16]. Similarly, re-configurability is un
a characteristic of a manufacturing system and may be explained as
of the system to quickly respond to the market variations (unexpected
through quick, effective and efficient configurations which fit optima
purposes [14].

Asree et al. [17] stated that the organizational culture and mana
tency have constructive association with responsiveness. Moreover
siveness results in higher business revenue. A suitable combination
bilities is vital for the current market where quick variations in
mix and demand are very common and where the requirement of
ery performance of a reliable manufacturing system is a basic ex
manufacturing operations, the overall direction for decision maki
the manufacturing strategies. Therefore, manufacturing strategy
weightage while focusing on responsiveness. Productivity and p
tives are the focal point for manufacturing strategy. The authors e
regards to responsiveness [19]. Holweg [20] established three con
siveness: first, the premise of responsiveness has a clear logic tha
array of production strategies. Second, these important factors
into three groups of responsiveness—volume, process and prod
tic understanding of responsiveness and its vital determinants
not be a single "holy grail" perception of how responsiveness ca
does a single approach applies to all segments.

From the literature it has been found that very little or no w
on productivity improvement with regards to responsiveness i
MSMEs. Thus, in this paper an attempt has been made to
of responsiveness on productivity of Indian MSMEs and to
between them.

11.2 Methodology

An attempt has been made in this investigation to establi
MSMEs responsiveness and their productivity, and hence
purpose of study, organizations considered are the micro
prises of North India in different manufacturing sectors. T
organizations are automobile components, leather produc
paper examines how responsive are MSMEs of North Ind
it for productivity improvement. The data were collecte
and semi-structured questionnaires. Thereafter, the acc
MSME executives were analyzed through statistical an

Based on the expert's advice on responsiveness an
items responsible for firm's productivity were identif
of respondents on responsiveness, statements on five-

Table 11.2 Reliability test

Reliability statistics		
Cronbach's alpha	Cronbach's alpha based or standardized items	No of items
0.863	0.869	7

328 responses were recorded and reliability test was conducted through Cronbach's alpha approach using SPSS 24.0 application.

The Cronbach's alpha for responsiveness was found to be 0.863, that is, internal consistency is good and acceptable. Therefore, we can conclude that the reliability of the factors is adequate and that the data can be used for further analysis (refer Table 11.2).

11.2.2 Descriptive Analysis

The data collected on each statement was then used to perform descriptive analysis. The results of the descriptive analysis are shown in Table 11.3.

Since the scale size was varying between 1 and 5, where 1 means completely disagree and 5 means completely agree, the mean score and standard deviation score in the table represents that majority of the respondents agreed that their organization responds rapidly and effectively to the unexpected demand changes and allocates resources optimally (Table 11.4).

Inter-item correlation

Producing a correlation matrix is an essential step to determine whether the study variables are related, and if they are, then to what extent. Table 11.5 represents the consolidate summary of the correlation among the independent variables used to generate inferences for responsiveness.

Table 11.3 Descriptive statistics

Item statistics			
	Mean	Std. deviation	N
RP1	4.42	0.782	328
RP2	3.62	0.852	328
RP3	4.24	0.998	328
RP4	3.99	1.006	328
RP5	4.18	1.039	328
RPG	4.31	0.898	328
RP7	3.87	0.960	328

Table 11.4 Summary of descriptive statistics

Summary item statistics

	Mean	Minimum	Maximum	Range	Maximum/ minimum	Variance	No. of items
Item means	4.091	3.622	4.424	0.802	1.221	0.078	7

Table 11.5 Inter-item correlation

Inter-item correlation matrix

	RP1	RP2	RP3	RP4	RP5	RP6	RP7
RP1	1.000	0.751	0.527	0.498	0.574	0.514	0.640
RP2	0.751	1.000	0.474	0.592	0.438	0.556	0.631
RP3	0.527	0.474	1.000	0.590	0.461	0.282	0.381
RP4	0.498	0.592	0.590	1.000	0.429	0.362	0.420
RP5	0.574	0.438	0.461	0.429	1.000	0.428	0.337
RP6	0.514	0.556	0.282	0.362	0.428	1.000	0.355
RP7	0.640	0.631	0.381	0.420	0.337	0.355	1.000

Since all cells are positive and are >0.3, that means there is a direct association among the study variables, that is, increase (decrease) in one variable will result in increase (decrease) in another and all variables are highly dependent on each other except RP3 and RP6. This means the change in "adjusting capacity to address demand changes" would result in "no" or "less" change in "effectively expedites emergency customer orders".

Improving Cronbach's alpha score

Sometimes to improve Cronbach's alpha (if <0.7), some items have to be deleted from the analysis. Cronbach's alpha statistics on deleted item was generated to see improvement in alpha score; however, in this case, it was not required as Cronbach's alpha score for responsiveness was 0.863. Also, the last column represents that the current alpha scores would not improve and will go down if we delete any item from the analysis. Thus, it was required to consider all the statements while performing test and analysis (Table 11.6).

χ^2-test

The χ^2-test for independence, also called Pearson's χ^2-test or the χ^2-test of association, was used to discover whether there is a relationship between productivity and responsiveness. On the basis of discussions with industry people, experts and review of previous studies, the following hypotheses were developed:

Table 11.6 Item statistics

Item-total statistics

	Scale mean if item deleted	Scale variance if item deleted	Corrected item-total correlation	Squared multiple correlation	Croribach's alpha if item deleted
RP1	24.21	17.844	0.791	0.690	0.827
RP2	25.02	17.477	0.770	0.682	0.827
RP3	24.40	17.677	0.596	0.439	0.850
RP4	24.65	17.324	0.638	0.484	0.844
RP5	24.45	17.570	0.578	0.403	0.853
RP6	24.33	18.729	0.533	0.361	0.857
RP7	24.77	17.940	0.593	0.466	0.850

H0 There exists no relationship between responsiveness and productivity, that is, responsiveness and productivity are independent.

H1 There exists a relationship between responsiveness and productivity improvement, that is, responsiveness and productivity are dependent.

Before conducting the test, we need to convert scale categories (completely disagree, disagree, neutral, agree and completely agree) to dichotomous categories (disagree and agree) to produce readable results for every statement. The dependent variable "productivity" was also converted to dichotomous category (increased and decreased) to see the impact of responsiveness on productivity (Tables 11.7 and 11.8).

11.3 Conclusion

- This study used samples of MSMEs of North India in sectors like automobile, leather and electrical manufacturing, with outstanding performance to investigate the impact of responsiveness on productivity and their performance.
- Considering the *p*-value (<0.05) and χ^2 value (>3.84) in the above analysis for all statements, we reject our null hypothesis and can safely conclude that there is statistically significant (more than 95%) relationship between responsiveness and productivity of MSMEs, that is, productivity and performance of MSMEs improve as they become more responsive.
- In addition, looking at the contingency coefficient and cross tabulation, we can conclude that the strength of association between responsiveness and productivity is good and responsiveness helps in productivity improvement.
- From the study it is also concluded that in the current competitive environment enterprises must become more responsive in their operations to enjoy benefits of the

Table 11.7 Cross-tabulation between responsiveness statements and productivity

Responsiveness statements		Productivity				
		Decreased		Increased		Total
		N	%	N	%	N
RP1	Disagree	6	75.0	2	25.0	8
	Agree	97	30.3	223	69.7	320
RP2	Disagree	24	82.8	5	17.2	29
	Agree	79	26.4	220	73.6	299
RP3	Disagree	19	86.4	3	13.6	22
	Agree	84	27.5	222	72.5	306
RP4	Disagree	17	81.0	4	19.0	21
	Agree	86	28.0	221	72.0	307
RP5	Disagree	17	85.0	3	15.0	20
	Agree	86	27.9	222	72.1	308
RP6	Disagree	13	76.5	4	23.5	17
	Agree	90	28.9	221	71.1	311
RP7	Disagree	24	82.8	5	17.2	29
	Agree	79	26.4	220	73.6	299

Table 11.8 Outcomes of χ^2-test

Statement	Pearson χ^2	d.f.	*p*-value	Valid cases	Contingency coefficient
RP1	7.235	1	0.007	328	0.147
RP2	38.951	1	0.000	328	0.326
RP3	33.069	1	0.000	328	0.303
RP4	25.572	1	0.000	328	0.269
RP5	28.403	1	0.000	328	0.282
RP6	16.906	1	0.000	328	0.221
RP7	38.951	1	0.000	328	0.326

Note The χ^2 value at 95% confidence level and 1 degree of freedom is 3.84

outcomes, like improving their productivity, performance and being competitive and to avoid being driven out of the market.

- The outcomes of the study can be applied in all manufacturing sectors of the country as from the investigation it has been found that in the current customer-driven markets and in the era of globalization, responsiveness of the enterprises has adequate impact on productivity and performance of the firms. That is more responsive enterprises are more productive and their performances also have improved.

References

1. Matson, J.B., McFarlane, D.C.: Assessing the responsiveness of existing production operations. Int. J. Oper. Prod. Manag. **19**(8), 765–784 (1999)
2. McFarlane, D., Chang, Y., Matson, J., Shaw, A.: A production responsiveness audit. In: Proceedings of POMS, San Francisco, USA (2001)
3. Ebrahim, Z., Ahmad, N.A., Muhamad, M.R.: Understanding responsiveness in manufacturing operations. Sci. Int. **26**(5) (2014)
4. Jimenez, C.H.O., Machuca, J.A., Garrido-Vega, P., Filippini, R.: The pursuit of responsiveness in production environments: from flexibility to reconfigurability. Int. J. Prod. Econ. **163**, 157–172 (2015)
5. Purvis, L., Gosling, J., Naim, M.M.: The development of a lean, agile and leagile supply network taxonomy based on differing types of flexibility. Int. J. Prod. Econ. **151**, 100–111 (2014)
6. Roh, J., Hong, P., Min, H.: Implementation of a responsive supply chain strategy in global complexity: the case of manufacturing firms. Int. J. Prod. Econ. **147**, 198–210 (2014)
7. Agarwal, R., Green, R., Brown, P.J., Tan, H., Randhawa, K.: Determinants of quality management practices: an empirical study of New Zealand manufacturing firms. Int. J. Prod. Econ. **142**, 130–145 (2013)
8. Fang, E.A., Wu, Q., Miao, C., Xia, J., Chen, D.: The impact of new product & operations technological practices on organization structure. Int. J. Prod. Econ. **145**(2), 733–742 (2013)
9. Liao, K., Deng, X., Marsillac, E.: Factors that influence Chinese automotive suppliers' mass customization capabilities. Int. J. Prod. Econ. **146**, 25–36 (2013)
10. da Silveira, G.J.C.: An empirical analysis of manufacturing competitive factors and offshoring. Int. J. Prod. Econ. **150**, 163–173 (2014)
11. Laurent Lim, L., Alpan, G., Penz, B.: Reconciling sales and operations management with distant suppliers in the automotive industry: a simulation approach. Int. J. Prod. Econ. **151**, 20–36 (2014)
12. Slack, N.: The flexibility of manufacturing systems. Int. J. Oper. Prod. Manag. **7**(4), 35–45 (1987)
13. Upton, D.M.: The management of manufacturing flexibility. Calif. Manag. Rev. **36**(2), 72–89 (1994)
14. Koren, Y.: General RMS characteristics. Comparison with dedicated and flexible systems. In: Reconfigurable Manufacturing Systems and Transformable Factories, pp. 27–45. Springer, Berlin, Heidelberg (2006)
15. Kalchschmidt, M., Nieto, Y., Reiner, G.: Managing demand through the enablers of flexibility: the impact of forecasting and process flow management. In: Rapid Modelling for Increasing Competitiveness: Tools and Mindset, pp. 265–276. Springer, London (2009)
16. Bernardes, E.S., Hanna, M.D.: A theoretical review of flexibility, agility and responsiveness in the operations management literature: toward a conceptual definition of customer responsiveness. Int. J. Oper. Prod. Manag. **29**(1), 30–53 (2009)
17. Asree, S., Zain, M., Rizal Razalli, M.: Influence of leadership competency and organizational culture on responsiveness and performance of firms. Int. J. Contemp. Hosp. Manag. **22**(4), 500–516 (2010)
18. McFarlane, D., Matson, J.: Assessing and improving the responsiveness of manufacturing production systems. In: IEE Seminar Customer Focused Manufacturing: Survival of the Fittest (1999). https://doi.org/10.1049/ic:19990802
19. Kritchanchai, D., MacCarthy, B.: Responsiveness and strategy in manufacturing. In: IEE Workshop on Responsiveness in Manufacturing (1998). https://doi.org/10.1049/ic:19980104
20. Holweg, M.: The three dimensions of responsiveness. Int. J. Oper. Prod. Manag. **25**(7), 603–622 (2005)

Chapter 12
Influence of Process Parameters on Weld Bead Geometry and Mechanical Properties in GTAW

Radha Kant, Shailendra Pandey, Renu Singh and Pranay Tanwar

Abstract In this paper, the structural and mechanical properties of IS:2062 mild steel after single-pass gas tungsten arc welding (GTAW) were investigated and evaluated, to reveal the weld strength and microhardness of welded joints. The properties of the welded joint are affected by various welding parameters, such as welding current, welding speed, gas flow rate and groove angle. This work concerns with mild steel plate joining with bevel angle (30°–50°) and also using filler material same as the base metal. The quality of welds will depend on the modeling of weld bead shape.

Keywords GTAW · Gas flow rate · Current · Microhardness and tensile strength

12.1 Introduction

Welding termed as fabrication process is used to join the materials, plastics and thermoplastics. The allied processes of welding technique, such as thermal cutting, brazing and soldering, provided the economical factor to the designer. A large number of welding and allied processes have come into industrial use in the last 30 years. Variations and extensions of these processes are being developed and put into practical use from time-to-time. Thus the metals join technology in a constant state of flux. The wide range and varieties of these processes enable the modern engineer to join almost all commercial metals and alloys in many different shapes and sizes and in thicknesses, ranging from a fraction of millimeter to over 500 mm.

Different ferrous materials, such as alloys of steel and many nonferrous metals and their alloys and metals such as zirconium, niobium and titanium, are extensively welded. Welding techniques play a vital role in the modern industry, such as TAPI pipe line, car, truck, bike assembly line, pressure vessels and power plants.

R. Kant (✉) · R. Singh · P. Tanwar
Delhi Technical Campus, Greater Noida, India
e-mail: mradhakant07@gmail.com

S. Pandey
Subharti Institute of Technology and Engineering, Meerut, India

© Springer Nature Singapore Pte Ltd. 2020
I. Singh et al. (eds.), *Trends in Manufacturing Processes*,
Lecture Notes on Multidisciplinary Industrial Engineering,
https://doi.org/10.1007/978-981-32-9099-0_12

Table 12.1 Summary of literature review

S. no.	Author	Work done/remark
1	Tseng et al.	The authors investigated the effect of TIG process using different flux, current, voltage and gap distance between the plates. The material used by the author is 316 L stainless steel
2	S.C. Juang et al.	The authors used Taguchi method to check the optimal condition of weld pool geometry
3	Ugur Esme et al.	The authors make 16 experimental orthogonal array and used Taguchi method to check the optimality of the TIG welding. The analysis of variance method (ANOVA) has been used to evaluate the characteristics of the weldment
4	R. Satish et al. (2012)	The authors applied ANN and fuzzy logic to investigate the optimality of the process parameter of the dissimilar pipe joints using GTAW
5	G. Haragopal et al. (2011)	The authors used aluminum alloy (AI-65032) to perform MIG welding and used Taguchi Method for the study of current and gas pressure
6	Qiu et al. (2011)	The authors studied about the microstructural characteristics and mechanical properties of the SiO_2 flux. He observed that the HAZ (heat-affected zone) in the joint is slightly wider than without using the flux
7	S. Krishnanunni et al.	Studied about the effect of welding condition on hardness of titanium material. He used TIG welding (argon as a shielding gas)
8	Indira Rani et al.	The authors used AA6351 material and performed GTAW/TIG welding with current 70–74 A, arc travel speed 700–760 mm/min and pulse frequency 3 and 7 Hz. He found that failure location of weldments occurred at HAZ

12.2 Literature Review

Many authors have done work to handle the issues related to single-pass gas tungsten arc welding (GTAW). Despite this, there has been very less work related to process parameters on weld bead geometry and mechanical properties in GTAW. Table 12.1 presents the summary of literature review.

12.3 Experimentation

The mild steel specimen of size 100 × 100 × 5 mm was cut from mild steel flat and mounted on shaper machine for V-groove machining providing bevel angle of 30° and 50°. The shaper and grinding machine located in the machine shop of central

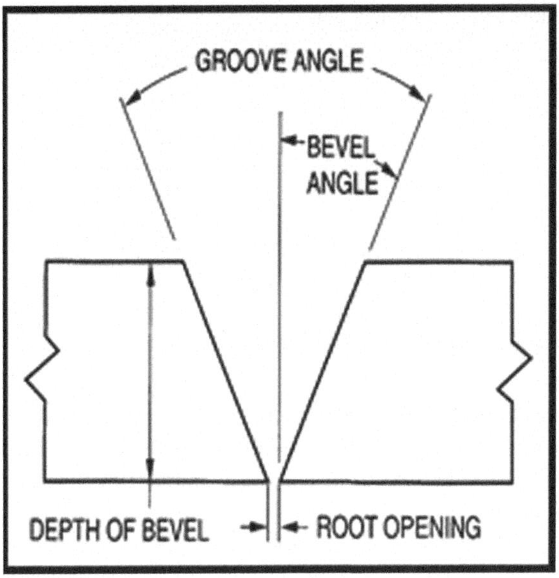

Fig. 12.1 Groove angle and bevel angle

workshop is used for the preparation of bevel angle. The shaper machine is having maximum travel of 18 in. Single-point cutting tool mounted on shaper machine tool post is used for machining. Grinding machine is used for the dressing of other sides of specimen for removing the burrs and sharp edges.

The number of specimen is according to fractional factorial design technique and is equal to 8 (since input variables are four and two levels of each factor). As number of factors are 4 and number of levels 2, so according to full-fractional design approach, total number of runs should be $2^4 = 16$. But the objective is to optimize the results that include cost also. Therefore, fractional factorial design of experiment method is used, which reduces the number of experiment to $2^{4-1} = 8$ and two numbers of replicates of design are used for analysis purpose (Fig. 12.1).

Sample Heading (Third Level). Only two levels of headings should be numbered. Lower level headings remain unnumbered; they are formatted as run-in headings.

12.3.1 Experiment Procedure

For finding the suitable factors and their levels for welding the specimen, experiments were performed. Table 12.2 shows the different process parameters at different levels. Travel speed, welding current, gas flow rate and bevel angle are the most important parameters. Similarly, factor levels were evaluated using the trial runs and subsequently inspecting the weld bead geometry and tensile strength.

Table 12.2 Parameters and levels

S. No.	Parameters	Low level (−1)	High level (+1)
1	Current (A)	110	150
2	Travel speed (mm/min)	60	90
3	Gas flow rate (L/min)	10	14
4	Bevel angle (degree)	30	50

12.4 Results and Analysis

Bead width is the width of the metal-deposited area on the plate over which the arc force is discharged. It is well known that the weld bead width is a major determinant of the weld quality, particularly in conditions where automatic equipments are used to fill up the grooves.

In fusion welding process the depth of weld penetration is generally recognized as the distance below the original surface of the work to which the molten metal progresses. Penetration is regarded as one of the most important dimensions of a welded joint as it determines the stress carrying capacity of a weld joint as well as its microstructure. The parameters used in the bead are shown in Table 12.3.

12.4.1 Analysis of Bead Width

The response factor, bead width, is measured and the values are tabulated in Table 12.4.

As the Cochran tabulated value is 0.68 and the experimental value is 0.43 which (experimental) is less than the tabulated value, it shows that the model is homogenous and the experiment fulfills the Cochran criteria. In Fig. 12.2 there is a design matrix for calculating the coefficient of bead width, and with the help of this table, model equation is developed for the bead width.

Table 12.3 Parameters and their values at two levels

Process parameter	Units	Type of parameter	Low level (−1)	High level (+1)
Current (A)	Ampere	Numeric	110	150
Travel speed (S)	mm/min	Numeric	60	90
Gas flow rate (F)	L/mm	Numeric	10	14
Bevel angle (β)	Degree	Numeric	30	50

Table 12.4 Experimental design matrix and measured responses of bead width

Trial no.	Obs W1	Obs W2	Obs W3	W avg.	ΔW	ΔW^2	$2\Delta W^2$
1	6.22	6.01	6	6.115	0.105	0.011025	0.02205
2	6.46	6.31	6.34	6.385	0.075	0.005625	0.01125
3	4.08	4.02	4.1	4.05	0.03	0.0009	0.0018
4	6.27	6.2	6.33	6.235	0.035	0.001225	0.00245
5	6.42	6.5	6.6	6.46	−0.04	0.0016	0.0032
6	7.41	7.5	7.52	7.455	−0.045	0.002025	0.00405
7	5.4	5.56	5.512	5.48	−0.08	0.0064	0.0128
8	6.65	6.35	6.561	6.5	0.15	0.0225	0.045
						0.0513	0.1026
						Tab val.	Homogeneous
	Cochran criteria			max/var∑	0.438596	0.68	Yes
	Variance of response			(Sq Si)/8	0.012825		

Further study shows that this model equation is nearly the same as estimated equation. Also, this model equation is the same as model equation developed with the help of design expert software.

Maximum penetration has been seen at higher welding current and lower weld travel speed, that is, 150 A and 60 mm/min, respectively. The increase in penetration as welding current is increased could be attributed to the enhanced arc force due to constriction of arc, and heat input per unit length of the weld bead resulted in higher current density that caused melting of a larger volume of the base metal and hence deeper penetration. In the present study, it has been found that lower weld travel speed of 60 mm/min results in better penetration.

12.5 Conclusion

From the analysis of experimental trials, it has been found that the weld bead geometry and mechanical properties (microhardness and tensile strength) have great influence of welding parameters, such as welding current, welding speed, gas flow rate and bevel angle on the weld joint.

- Current is the most effective parameter for bead width followed by arc travel speed. Gas flow rate and bevel angle have very less significance of effectiveness. Minimum optimized value bead width is desired.
- Travel speed is the most effective parameter for depth of penetration followed by current. Gas flow rate has very less significance of effectiveness and bevel angle is nearly insignificant. Larger value of depth of penetration is desired.

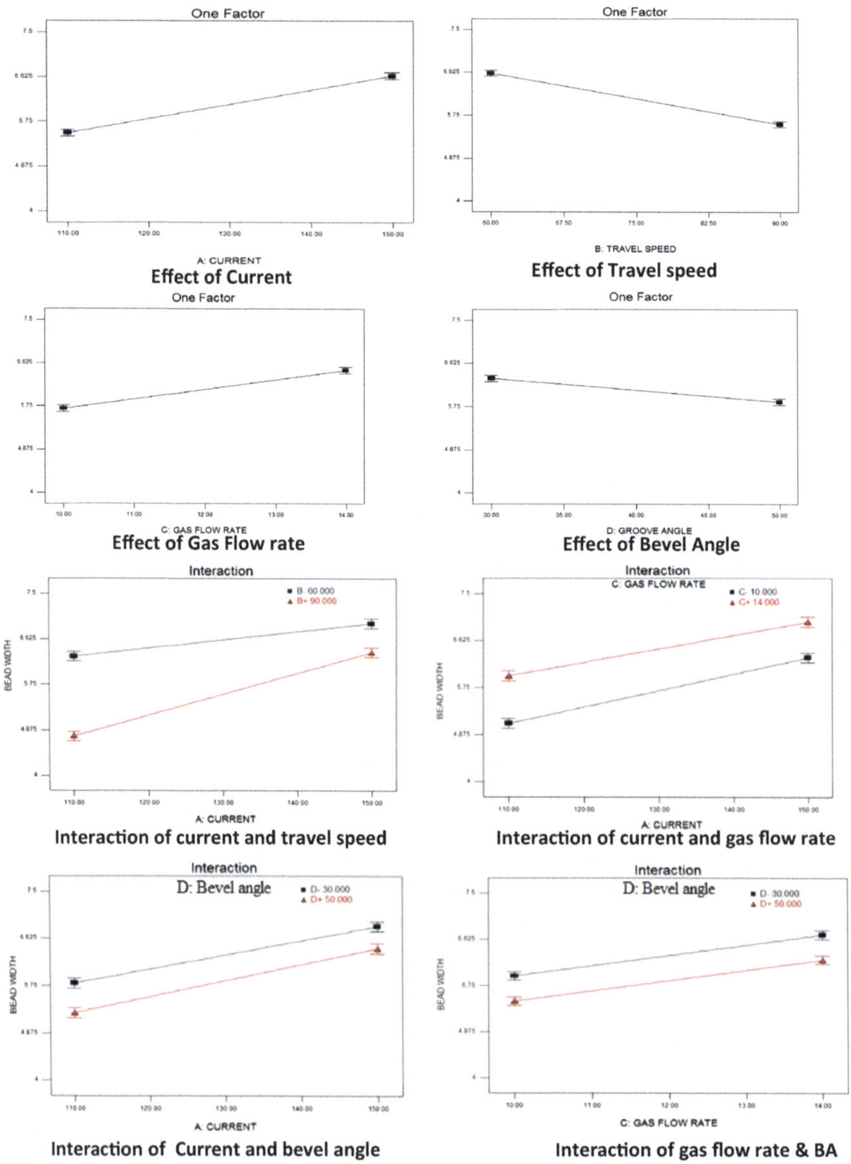

Fig. 12.2 Effects of process parameter on bead width

- Maximum tensile strength of 406.2 N/mm^2 was observed at weld speed of 60 mm/min (for 30° bevel angle and 150 A current).
- At lower welding speeds, strength is more due to more intensity of current. This indicates that lower range of weld speed is suitable for achieving maximum tensile strength.

It is observed that the hardness is higher in the weldment and HAZ than the base metal. Hardness value of the weld zone changes with the distance from weld centre due to change of microstructure.

Bibliography

1. Nadkarni, S.: Modern Arc Welding Technology. IBH Publication, New Delhi, India (1988)
2. Cary, H.B., Helzer, S.C.: Modern Welding Technology, pp. 246–49. Pearson Education, Upper Saddle River, New Jersey (2005)
3. Weman and Klas.: Welding Processes Handbook, p. 31. CRC Press LLC, New York, NY (2003)
4. Weman and Klas.: Welding Processes Handbook, pp. 60–62. CRC Press LLC, New York, NY (2003)
5. Shoeb, M., Parvez, M., Kumari, P.: Effect of MIG welding input process parameters on weld bead geometry on HSLA steel. Technology (IJEST) **5**(01), 200–212 (2013)
6. Peasuraa, P., Watanapab, A.: Influence of shielding gas on aluminium alloy 5083 in gas tungsten arc welding. In: 2012 International Workshop on Information and Electronics Engineering (IWIEE), Procedia Engineering 29, pp. 2465–2469 (2012)
7. Kumar, V., Singh, G., Yusufzai, M.Z.K.: Effect of process parameters of gas metal arc welding on dilution in cladding of stainless steel on mild steel. MIT Int. J. Mech. Eng. **2**(2), 127–131 (2012)
8. Haragopal, G., Ravindra Reddy, P.V.R., Chandra Mohan Reddy, G., Subrahmanyam, J.V.: Parametric design for MIG welding of Al-65032 alloy using Taguchi Technique. J. Sci. Ind. Res. **70**, 844–858 (2011)
9. Palani, P.K., Saju, M.: Modelling and optimization of process parameters for TIG welding of aluminium-65032 using response surface methodology. Int. J. Eng. Res. Appl. **3**(2), 230–236 (2013)
10. Narang, H.K., Singh, U.P., Mahapatra, M.M., Jha, P.K.: Prediction of the weld pool geometry of TIG arc welding by using fuzzy logic controller. Int. J. Eng. Sci. Technol. **3**(9), 77–85 (2012)
11. Nagesh, D.S., Datta, G.L.: Genetic algorithm for optimization of welding variables for height to width ratio and application of ANN for prediction of weld bead geometry for TIG welding process. Appl. Soft Comput. **10**, 897–907 (2010)
12. Ador Welding Limited.: Modern Arc Welding Technology. Oxford & IBH Publishing Co. Pvt. Ltd.
13. Parmar, R.S.: Welding Processes and Technology. Khanna Publishers
14. Khanna, O.P.: A Textbook of Welding Technology. Dhanpat Rai Publications
15. Welding Handbook Volume 2, "Welding Processes" by American Welding Society (AWS)

Chapter 13
Supplier Selection Using Fuzzy-AHP: A Case Study

Narayan Agrawal and Shashi Kant

Abstract Application of multi-criteria decision-making to resolve industrial problems is well-known. The present research is related to the use of fuzzy-AHP in selection of a supplier in the presence of conflicting criteria. There is always an uncertainty present if the crisp values are used. To avoid uncertainty, in this study linguistic variables have been used. An integrated model of DEMATEL and FAHP has been presented here. The weights obtained from the DEMATAL have been used to calculate the ranking of suppliers through FAHP. The rankings obtained through FAHP have also been compared with the rankings calculated through FTOPSIS [13] for the purpose of validation.

Keywords MCDM · Fuzzy-AHP · Decision matrix

13.1 Introduction

In the presence of conflicting criteria, it is always difficult to take the right and logical decision by the management. The trade-off between criteria is always required to optimize the recourses and selection of good alternative. Thus, the goals of this study are: (1) to use weights/ranking criteria obtained through DEMATEL (refer to one of the unpublished and communicated manuscript of the same authors); (2) selecting the optimal supplier using FAHP; and (3) verification/validation of the result obtained from FAHP through FTOPSIS [1].

In the past research, FAHP has been used to resolve many real and practical problems. Sirisawat et al. [2] proposed a methodology using fuzzy-AHP and fuzzy-TOPSIS for the final ranking of the solutions of reverse logistics implementation. They have also considered Thailand's electronics industry to present the

N. Agrawal (✉)
Delhi Institute of Tool Engineering, Okhla Ph-2, Delhi, India
e-mail: n_nsit@yahoo.co.in

S. Kant
Department of Mechanical Engineering, Manav Rachna International Institute of Research and Studies, Faridabad, India

© Springer Nature Singapore Pte Ltd. 2020
I. Singh et al. (eds.), *Trends in Manufacturing Processes*,
Lecture Notes on Multidisciplinary Industrial Engineering,
https://doi.org/10.1007/978-981-32-9099-0_13

methodology. Sensitivity analysis has also been carried out to show the change in priority after changing weight of criteria [2]. Dožić et al. proposed a methodology for choosing aircraft type to meet the market conditions and airline's requirements for known route networks. They have used FAHP and LFPP for solving this selection problem [3].

Ilbahar et al. presented a novel approach to risk assessment for occupational health and safety using Pythagorean fuzzy-AHP and fuzzy inference system [4]. Goyal et al. discussed the disadvantages of using extent analysis method and proposed a utility-based non-linear fuzzy-AHP optimization model for the selection of network in heterogeneous wireless networks [5]. To estimate probability of success and risk factors that affect EER, Ping et al. proposed a model using FAHP and decomposition method [6]. Jayawickrama et al. used fuzzy logic to convert the qualitative judgments into evaluable numbers and presented a generic model to evaluate manufacturing plant sustainability using fuzzy analytic hieratical process (AHP). They have also discovered a tool that can perform analysis at variable resolutions of available plant data [7]. Fuzzy logic and fuzzy analytical hierarchy process (AHP) were used for data processing by Zhang et al. [8] and they found that fuzzy logic and AHP were performed for porphyry and hydrothermal vein copper deposits in the Dananhu-Tousuquan island arc, Xinjiang. The results of the two methods have also been compared by the authors [8]. Li Wenhua et al. proposed a hybrid approach using fuzzy-AHP and two-tuple fuzzy linguistic approach to examine in-flight service quality. They also demonstrated feasibility study of the comprehensive performance of three airlines' in-flight service quality in China [9].

A new multi-expert fuzzy method integrating intuitionist fuzzy data envelopment analysis (DEA) and intuitionist fuzzy analytic hierarchy process (IF-AHP) has been proposed by Otay et al. for performance evaluation of healthcare institutions [10]. Celik et al. developed a comprehensive MCDM method extended by interval type-2 fuzzy sets (IT2FSs), for selecting appropriate shiploader type in maritime transportation. They also incorporated AHP and TOPSIS under IT2FSs environment to overcome more of the uncertainty of experts' judgment and expression in decision-making [11]. Chena et al. proposed a new approach for evaluating teaching performance using combination of fuzzy-AHP and fuzzy comprehensive evaluation method [12]. Akkaya et al. prepared a questionnaire to conduct a study on 60 students who study at Industrial Engineering Department at different universities in Turkey. Fuzzy-AHP and fuzzy-MOORA (multi-objective optimization by ratio analysis) have been used. The study was done for seven sectors and ten criteria [13].

On the basis of past literature, it was found that the integrated approach has not been applied. Especially, use of weights determined from DEMATEL as input to the fuzzy-AHP and result further compared the result obtained from fuzzy-TOPSIS for validation of the model. This integrated approach gives the rankings of the criteria and the suppliers. The organization of the paper is as follows: Sect. 13.2 discusses the integrated model. Section 13.3 provides the supplier through fuzzy AHP, and Sect. 13.4 presents results and discussion and conclusion.

13.2 Integrated Model

The proposed integrated model has been shown through flowchart in Fig. 13.1, and the linguistic variables are shown in Table 13.1.

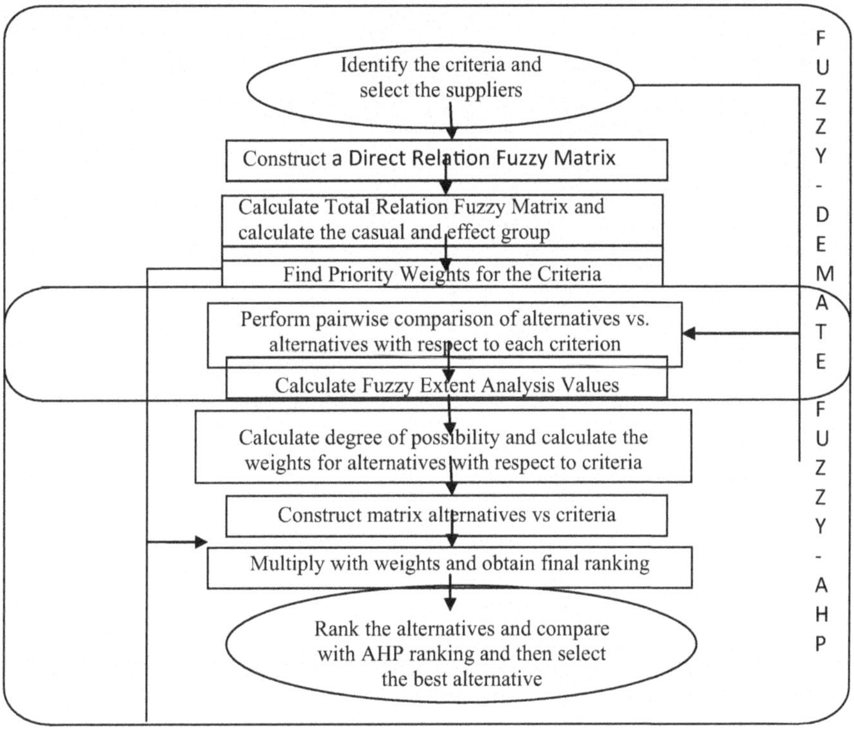

Fig. 13.1 Flowchart for ranking through FAHP

Table 13.1 Linguistic variables and TFN

Linguistic variables	Triangular fuzzy numbers
Very low (VL)	(1, 1, 1)
Low (L)	(2, 3, 4)
Good (G)	(4, 5, 6)
High (H)	(6, 7, 8)
Very high (VH)	(8, 9, 9)

13.2.1 Fuzzy-Ahp Model

Stage-1: Weight from Fuzzy-DEMATEL (refer to unpublished and communicated manuscript of same authors)

$$w_i = \frac{W_i}{\sum_{i=1}^{n} W_i} \tag{13.1}$$

$$W_i = \left\{ \left(\check{D}_i^{def} + \check{R}_i^{def} \right)^2 + (\check{D}_i^{def} - \check{R}_i^{def})^2 \right\}^{\frac{1}{2}} \tag{13.2}$$

$$\check{D}_i = \left[\sum_{j=1}^{n} t_{ij} \right] (i = 1, 2, \ldots, n) \tag{13.3}$$

$$\check{R}_i = \left[\sum_{j=1}^{n} t_{ij} \right] (j = 1, 2, \ldots, n) \tag{13.4}$$

$$T_{ij} = \lim_{w \to \infty} (A_{ij} + A_{ij}^2 + \cdots + A_{ij}^w) = A_{ij}(I - A_{ij})^{-1} \tag{13.5}$$

where I denotes the identity matrix and

$$t_{ij} = \left[t_{ij,l}, t_{ij,m}, t_{ij,u} \right].$$

Stage-2: Fuzzy Extension Analysis

Step 1. Construct a decision matrix by using TFN and the TFN's arithmetic operational law, which is given by:

$$(l_i, m_i, u_i)^{-1} \approx (1/u_i, 1/m_i, 1/l_i) \quad \text{where } i = 1, 2, \ldots, n \tag{13.6}$$

Step 2. Now normalize the matrix. For normalization check whether $l_1 > u_2 = 0$ or not.

If $(l_1 - u_2 > 0)$, then,

$$\left(\frac{l_i}{l_i + m_i + u_i}, \frac{m_i}{l_i + m_i + u_i}, \frac{u_i}{l_i + m_i + u_i} \right) \tag{13.7}$$

for all the elements, and if $(l_1 - u_2 < 0)$, then use other arithmetic extension laws of TFN.

Step 3. Find the value of fuzzy synthetic extent with respect to the ith object.

$$S_i = \sum_{j=1}^{m} M_i^j \odot \left[\sum_{j=1}^{n} \sum_{j=1}^{m} M_i^j \right]^{-1} \tag{13.8}$$

Step 4. Define the degree of possibility between two fuzzy numbers.

$$v(S_m \geq S_n) = hgt(S_m \cap S_n) = \mu_{Sm}(d) = \begin{cases} 1, & m_m > m_n \\ \frac{(l_n - u_m)}{(m_m - u_m) - (m_n - l_n)}, & m_m > m_n \end{cases} \tag{13.9}$$

where d is the ordinate of the highest intersection point between μ_{Sm} and μ_{Sn}.

Step 5. The degree of possibility for a convex number to be greater than p convex fuzzy $S_i (i = 1, 2, \ldots, p)$ numbers are expressed as:

$$\begin{aligned} v\left(S \geq S_1, S_2, \ldots S_p\right) &= v\left[(S \geq S_1), (S \geq S_2), \ldots\ldots, \left(S \geq S_p\right)\right] \\ &= \min v(S \geq S_i) = d'(C_i), \quad i = 1, 2, \ldots p \end{aligned} \tag{13.10}$$

***Step* 6**. Calculate the weight vector for alternative with respect to each criterion.

$$W' = \left(d'(C_1), d'(C_2), \ldots, d'(C_n)\right)^T \tag{13.11}$$

Normalize this value through Eq. 13.8 and obtain the final weight.

$$w' = (d(C_1), d(C_2), \ldots, d(C_n))^T \tag{13.12}$$

The above integrated model has been applied to an industry to find out best supplier with the five criteria, that is, *delivery time*, *cost*, *product quality*, *responsiveness* (support and feedback, adjustment to client) and *credit term* (payment adjustment) and five supplier named as SA, SB, SC, SD and SE. The detailed computational work of the proposed approach is shown in the following section. However, after using fuzzy-DEMATEL approach (unpublished but communicated research paper from the same authors) for the identification of significant criteria, insignificant criteria, that is, product quality has been removed from the further study for the selection of suppliers through fuzzy-AHP and fuzzy-TOSIS, however, it has also been verified from the calculations before removal of the mentioned criteria that inclusion or exclusion of this criteria having no impact on selection. Thus only four criteria have been considered here.

13.3 Supplier Selection Through Fuzzy-AHP with Extension Analysis

Now by performing a pairwise comparison between alternatives vs. alternatives with respect to the given criteria, decision matrix is constructed by the expert. These matrices are solved by the fuzzy extension analysis to obtain the weights of alternatives with respect to certain criterion. Matrix with respect to criterion delivery time (DT) is given in Tables 13.2 and 13.3.

Now, Eq. 13.10 is used for calculating the value of fuzzy extent analysis.

$S_{DT1} = (1.3746, 1.6667, 1.9587) \times (9.7408, 8.2274, 7.0318)^{-1} = (0.1411, 0.2026, 0.2786)$

$S_{DT2} = (1.4446, 1.6705, 1.8849) \times (9.7408, 8.2274, 7.0318)^{-1} = (0.1483, 0.2030, 0.2681)$

$S_{DT3} = (1.4437, 1.6531, 1.9033) \times (9.7408, 8.2274, 7.0318)^{-1} = (0.1482, 0.2009, 0.2707)$

Table 13.2 Matrix with respect to DT

Supplier	SA	SB	SC	SD	SE
SA	(1, 1, 1)	(4, 5, 6)	(6, 7, 8)	(2, 3, 4)	(4, 5, 6)
SB	(0.16, 0.20, 0.25)	(1, 1, 1)	(4, 5, 6)	(8, 9, 9)	(4, 5, 6)
SC	(0.12, 0.14, 0.16)	(0.16, 0.20, 0.25)	(1, 1, 1)	(4, 5, 6)	(6, 7, 8)
SD	(0.25, 0.33, 0.50)	(0.11, 0.11, 0.12)	(0.16, 0.20, 0.25)	(1, 1, 1)	(2, 3, 4)
SE	(0.16, 0.20, 0.25)	(0.16, 0.20, 0.25)	(0.12, 0.14, 0.16)	(0.25, 0.33, 0.50)	(1, 1, 1)

Table 13.3 Linguistic variables and TFN

Supplier	SA	SB	SC	SD	SE	W_{DTs}
SA	(0.33, 0.33, 0.33)	(0.26, 0.33, 0.40)	(0.28, 0.33, 0.38)	(0.22, 0.33, 0.44)	(0.26, 0.33, 0.40)	0.2037
SB	(0.27, 0.32, 0.40)	(0.33, 0.33, 0.33)	(0.26, 0.33, 0.40)	(0.30, 0.34, 0.34)	(0.26, 0.33, 0.40)	0.2045
SC	(0.28, 0.32, 0.38)	(0.27, 0.32, 0.40)	(0.33, 0.33 0.33)	(0.26, 0.33, 0.40)	(0.28, 0.33, 0.38)	0.2010
SD	(0.23, 0.30, 0.46)	(0.32, 0.32, 0.36)	(0.27, 0.32, 0.40)	(0.33, 0.33, 0.33)	(0.22, 0.33, 0.44)	0.1955
SE	(0.27, 0.32, 0.40)	(0.27, 0.32, 0.40)	(0.28, 0.32, 0.30)	(0.23, 0.30, 0.46)	(0.33, 0.33, 0.33)	0.1953

$S_{DT4} = (1.3766, 1.6187, 2.0047) \times (9.7408, 8.2274, 7.0318)^{-1} = (0.1413, 0.1967, 0.2851)$

$S_{DT5} = (1.3923, 1.6184, 1.9892) \times (9.7408, 8.2274, 7.0318)^{-1} = (0.1429, 0.1967, 0.2859)$.

Degree of possibility between two fuzzy numbers and convex fuzzy numbers are obtained from the Eqs. 13.11 and 13.12, respectively.

$v(S_1 \geq S_2) = 0.9965$, $v(S_1 \geq S_3) = 1$, $v(S_1 \geq S_4) = 1$, $v(S_1 \geq S_5) = 1$,
$v(S_2 \geq S_1) = 1$, $v(S_2 \geq S_3) = 1$, $v(S_2 \geq S_4) = 1$, $v(S_2 \geq S_5) = 1$,
$v(S_3 \geq S_1) = 0.9874$, $v(S_3 \geq S_2) = 0.9830$, $v(S_3 \geq S_4) = 1$, $v(S_3 \geq S_5) = 1$
$v(S_4 \geq S_1) = 0.9611$, $v(S_4 \geq S_2) = 0.9560$, $v(S_4 \geq S_3) = 0.9704$, $v(S_4 \geq S_5) = 1$
$v(S_5 \geq S_1) = 0.9603$, $v(S_5 \geq S_2) = 0.9551$, $v(S_5 \geq S_3) = 0.9697$, $v(S_5 \geq S_4) = 0.9998$.

$d(C_1) = v(S_1 \geq S_2, S_3, S_4, S_5) = \min(0.9965, 1, 1, 1) = 0.9965$
$d(C_2) = v(S_2 \geq S_1, S_3, S_4, S_5) = \min(1, 1, 1, 1) = 1$
$d(C_3) = v(S_3 \geq S_1, S_2, S_4, S_5) = \min(0.9874, 0.9830, 1, 1) = 0.9830$
$d(C_4) = v(S_4 \geq S_1, S_2, S_3, S_5) = \min(0.9611, 0.9560, 0.9704, 1) = 0.9560$
$d(C_5) = v(S_5 \geq S_1, S_2, S_3, S_4) = \min(0.9603, 0.9551, 0.9697, 0.9998) = 0.9551$.

Now by using the above-mentioned equations, priority weight for alternatives with respect to criterion DT is as follows:

$W_{DT} = (0.9965, 1, 0.9830, 0.9560, 0.9551)$
$w_{DT} = (0.2037, 0.2045, 0.2010, 0.1955, 0.1953)$.

Similarly, priority weights for other criteria have also been calculated as shown in Table 13.4. Now all the priorities weights obtained from the pairwise comparison of the alternatives with respect to the criteria are mapped in Table 13.4 along with the priority weight (w_i) obtained from the fuzzy-DEMATEL method. Now the product sum is calculated to find the final ranking of the alternatives with respect to the criteria.

Table 13.4 Final ranking (FAHP)

Criteria	DT	Cost	Resp.	CT	Weight
SA	0.2037	0.2034	0.2065	0.2062	0.2047
SB	0.2045	0.2034	0.2020	0.2022	0.2034
SC	0.2010	0.1969	0.2016	0.2009	0.1999
SD	0.1955	0.1991	0.1950	0.1953	0.1970
SE	0.1953	0.1972	0.1948	0.1954	0.1951
W_i	0.2581	0.1521	0.1798	0.1734	

Table 13.5 Ranking comparison

Fuzzy-AHP	SA > SB > SC > SD > SE
Fuzzy-TOPSIS	SA > SC > SB > SD > SE

13.4 Results and Discussion

The priority weights shown in Table 13.4 in the last column provide the final ranking to the suppliers. According to the obtained final weights, supplier A (SA) is found to be the best supplier for supplying the metal sheets to the company and the supplier E (SE) is found to be the least possible alternatives with least value. The ranking order for suppliers is as follows: SA > SB > SC > SD > SE. Furthermore, when the same problem solved through fuzzy-TOPSIS, supplier A is found to be the best supplier for the chosen business firm from both the methods. Suppliers B and C are interchangeably good as per the MCDM methods.

Suppliers D and E are the least desirable suppliers. Table 13.5 shows the complete ranking obtained from fuzzy-AHP and fuzzy-TOPSIS. In fuzzy-AHP ranking, supplier B is the second possible alternative, whereas in fuzzy-TOPSIS, it is found that supplier C is the second possible alternative. If there is very slight difference between crisp values of weights and decision matrix elements values, then it is possible that ranking obtained from two different MCDM methods may vary. It also occurs in this case that suppliers B and C's positions interchanged with each other as shown in Table 13.5. But it is also clear that the best three suppliers and least two suppliers are the same. For further study, other MCDM methods such as ANP, ELECTRE, VIKOR and PROMETHEE can also be integrated.

References

1. Agrawal, N., Shashikant: Integrated fuzzy-dematel, fuzzy topsis approach for supplier selection: a case study. Adv. Res. Electr. Electron. Eng. (AREEE) **5**(2), 129-133 (2018)
2. Sirisawat, P., Kiatcharoenpol, T.: Fuzzy AHP-TOPSIS approaches to prioritizing solutions for reverse logistics barriers. Comput. Ind. Eng. **117**, 303–318 (2018)
3. Dožić, S., Lutovac, T., Kalić, M.: Fuzzy AHP approach to passenger aircraft type selection. J. Air Trans. Manag. **68**, 165–175 (2018)
4. Ilbahar, E., Karaşan, A., Cebi, S., Kahraman, C.: A novel approach to risk assessment for occupational health and safety using Pythagorean fuzzy AHP & fuzzy inference system. Saf. Sci. **103**, 124–136 (2018)
5. Goyal, R.K., Kaushal, S., Sangaiah, A.K.: The utility based non-linear fuzzy AHP optimization model for network selection in heterogeneous wireless networks. Appl. Soft Comput. **67**, 800–811 (2018)
6. Ping, P., Wang, K., Kong, D., Chen, G.: Estimating probability of success of escape, evacuation, and rescue (EER) on the offshore platform by integrating Bayesian network and fuzzy AHP. J. Loss Prev. Process Ind. **54**, 57–68 (2018)
7. Jayawickrama, H.M.M.M., Kulatunga, A.K., Mathavan, S.: Fuzzy AHP based plant sustainability evaluation method. Procedia Manuf. **8**, 571–578 (2017)

8. Zhang, N., Zhou, K., Du, X.: Application of fuzzy logic and fuzzy AHP to mineral prospectivity mapping of porphyry and hydrothermal vein copper deposits in the Dananhu-Tousuquan island arc, Xinjiang, NW China. J. Afr. Earth Sci. **128**, 84–96 (2017)
9. Li, W., Yu, S., Pei, H., Zhao, C., Tian, B.: A hybrid approach based on fuzzy AHP and 2-tuple fuzzy linguistic method for evaluation in-flight service quality. J. Air Trans. Manag. **60**, 49–64 (2017)
10. Otay, İ., Oztaysi, B., Onar, S.C., Kahraman, C.: Multi-expert performance evaluation of healthcare institutions using an integrated intuitionistic fuzzy AHP&DEA methodology. Knowl. Based Syst. **133**, 90–106 (2017)
11. Celik, E., Akyuz, E.: An interval type-2 fuzzy AHP and TOPSIS methods for decision-making problems in maritime transportation engineering: the case of ship loader. Ocean Eng. **155**(1), 371–381 (2018)
12. Chena, J.-F., Hsieh, H.-N., Do, Q.H.: Evaluating teaching performance based on fuzzy AHP and comprehensive evaluation approach. Appl. Soft Comput. **28**, 100–108 (2015)
13. Akkaya, G., Turanoğlu, B., Öztaş, S.: An integrated fuzzy AHP and fuzzy MOORA approach to the problem of industrial engineering sector choosing. Expert Syst. Appl. **42**(24), 9565–9573 (2015)

Chapter 14
Friction Stir Welding of AA 8011: Mechanical and Microstructure Behavior

Mohd Atif Wahid, Nidhi Sharma, Pankul Goel, Aabir Mohsin and Ankit Singh

Abstract This paper presents the evaluation of the mechanical properties and microstructure of the butt joint of AA 8011 produced by friction stir welding (FSW). The effect of the important FSW parameters on ultimate tensile strength (UTS) was determined using Taguchi L9 orthogonal arrays (OA) for analyzing the main effects and variance. From the results of the study, it was found that the most dominating factor affecting the UTS is welding speed, contributing 61.87%, followed by shoulder diameter with 18.02% and lastly tool rotational speed with 13.38%. The maximum UTS of 142.1 MPa higher than the base metal (BM) strength was attained at ideal settings of rotational speed at 560 rpm, welding speed of 100 mm/min and tool shoulder of 16 mm. The microstructure revealed dynamically recrystallized fine and equiaxed grains in SZ. The microhardness in SZ was found to be greater than the BM/TMAZ/HAZ.

Keywords Aluminum · Friction stir welding · Mechanical properties · Microstructure · Ultimate tensile strength · Taguchi method · Optimization

M. A. Wahid (✉)
Mechanical and Automation Engineering Department, Delhi Technical Campus, Greater Noida, India
e-mail: wahidatif89@gmail.com

N. Sharma
Mechanical Engineering Department, Greater Noida Institute of Technology, Greater Noida, India
e-mail: nid.sharma83@gmail.com

P. Goel
Department of Mechanical Engineering, Jamia Millia Islamia, New Delhi, India
e-mail: pankul.goel@gmail.com

A. Mohsin · A. Singh
Department of Mechanical Engg, JSS Academy, Noida, India
e-mail: aabirmohsin@gmail.com

© Springer Nature Singapore Pte Ltd. 2020

I. Singh et al. (eds.), *Trends in Manufacturing Processes*,
Lecture Notes on Multidisciplinary Industrial Engineering,
https://doi.org/10.1007/978-981-32-9099-0_14

14.1 Introduction

Ever since the FSW was invented, it is commonly used for joining aluminum alloys (AAs). Unlike other fusion welding (FW) methods, the FSW does not require consumable electrode and protecting gases [1–3]. The major advantage of using FSW above many other FW practices is that the material that is being welded does not undergo melting. Owing to this property, many metallurgical problems present in FW processes, like spatter, shrinkage, distortion and porosity, can be prevented [1]. In FSW, the tool is plunged into the BM and navigated along the abutting edges of the joint. The frictional heating between the tool and the BM along with extreme plastic deformation of the base material (BM) causes joining of different pieces together [4–6], as shown in Fig. 14.1. The welding of these pieces is performed below its solidus temperature, eradicating quite a lot of FW defects, thus offering exceptional weld strength and ductility. FSW is fully accomplished to offer outstanding prospects for novel structural designing notions in numerous industries, like marine, aerospace, automobile and railway.

The FSW process is tremendously affected by the thermal cycles [7]. The welding of materials with FSW causes dissolution or coarsening of the strengthening precipitates due to thermal cycles established during the process, thus leading to deterioration in the mechanical performance of the joints [6]. For achieving good mechanical properties and defect-free joining, proper selection of different FSW parameters affecting the heat generation is required [8]. Various optimization approaches are used today for prediction purposes [5]. Taguchi technique is one of the prominent optimization technique used for predicting the mechanical properties of the FSW joint [8]. Lakshminarayan et al. [9] performed experiment on the AA 7039 and determined the consequence of FSW parameter on the UTS. Jayaraman et al. [10] investigated and concluded that the welding speed is the dominating factor affecting the UTS. Wahid et al. [11] optimized the FSW parameters for maximum UTS, both in air and water cooling media. They obtained increased UTS in water in comparison

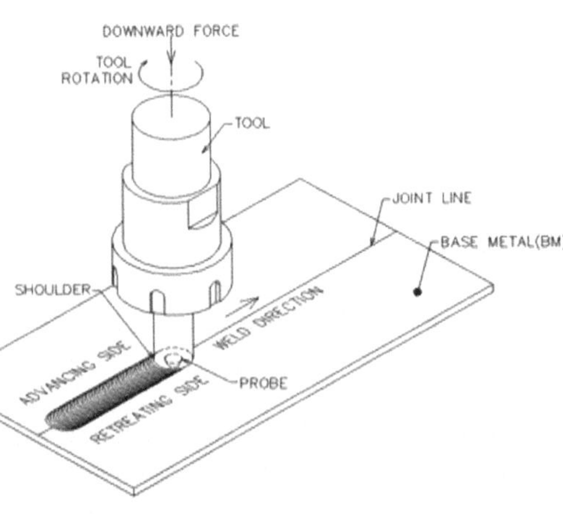

Fig. 14.1 Schematic representation of FSW

to air due to reduction in heat generation because of water cooling. Rajakumar et al. [12] investigated FSW joints of different AAs using Taguchi methodology.

In this paper, the optimal FSW condition for obtaining the maximum UTS for AA 8011 is carried out using Taguchi method. The optimal condition for the welding is determined based on three input constraints, which are shoulder diameter, rotational speed and welding speed of the tool. The paper also presents the results of microstructural and microhardness investigations of the welds achieved at the optimum conditions for friction stir welded joint.

14.2 Experimental Procedure

14.2.1 Joint Fabrication

The material that is used in this experiment is AA 8011, having the dimensions 200 × 50 × 3 mm. Low weight, corrosion resistance and easy shaping of the finished product make this alloy favorable for various applications. The composition of AA 8011 is given in Table 14.1. All the experimentations were performed on vertical milling machine dedicated to perform FSW (Fig. 14.2) with rigid fixture for suitable clamping of plates. The tool having material H13 was prepared with three different shoulder diameters of 14, 16 and 18 mm. Tool pin diameter, pin length and tilt angle of 6 mm, 2.70 mm and 1.5°, respectively, were used.

Table 14.1 The alloy composition of the BM (wt%)

Mg	Mn	Cr	Si	Fe	Zn	Ti	Cu	Al
0.011	0.10	0.002	0.81	0.61	0.001	<0.005	0.018	Balance

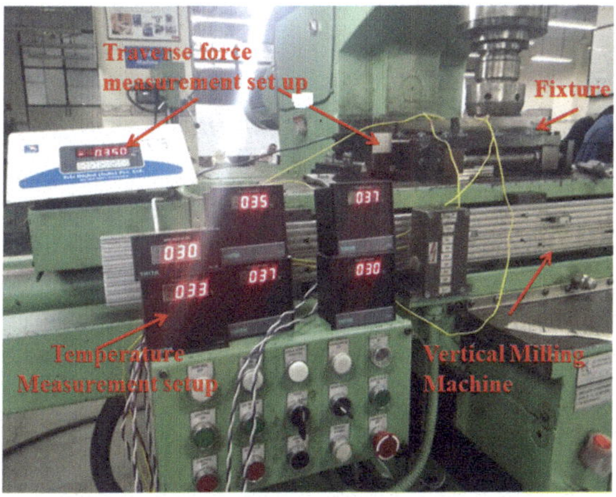

Fig. 14.2 FSW arrangement with fixture and other auxiliary units

Table 14.2 FSW parameters and its levels

Parameters	Symbol	Level 1	Level 2	Level 3
Shoulder diameter (mm)	A	14	16	18
Rotational speed (rpm)	B	560	710	900
Welding speed (mm/min)	C	63	80	100

Table 14.3 Design matrix (L9 OA)

Run	A	B	C	UTS (MPa)	S/N ratio
1	1	1	1	90.7	39.15
2	1	2	2	107	40.59
3	1	3	3	129.6	42.25
4	2	1	3	122.7	41.78
5	2	2	1	116.3	41.31
6	2	3	2	120.9	41.65
7	3	1	2	116.8	41.35
8	3	2	3	142.1	43.05
9	3	3	1	106.4	40.53

14.2.2 Determining the Working Levels

Various numbers of pre-experiments (trial runs) were performed at different parameters before carrying out the final experiments. These experiments were performed in order to predict the possible working range and levels. The final working levels were decided by tensile test of butt joint and visual inspection of cross section of welded joint (must be free of any visual defects). The chosen parameters along with the range are revealed in Table 14.2. The design matrix conferring Taguchi L9 OA is given in Table 14.3.

14.2.3 Testing of the Response Variables

Wire EDM machine was incorporated for cutting the tensile, microhardness and metallographic test pieces. The joints UTS in the transverse direction was estimated by a tensometer at a cross head feed of 2 mm/min. Microstructure samples were polished and etched with the help of Keller's etchant and inspected by a metallurgical microscope. Microhardness was evaluated using hardness testing machine at the joint mid-section by indenting a load of 2 N at 0.5 mm indentation distance and delay time of 20 s.

14.3 Results and Discussion

14.3.1 Tensile Testing

Defect-free FSW of AA 8011 was successfully performed at different parametric combination. The data were analyzed using standard statistical software Minitab. The criterion, larger the better, was chosen for UTS, as joints with the highest strength are desirable. The experimental data and the S/N ratio calculated for the UTS are shown in Table 14.4.

From the analysis of mean (ANOM) shown in Table 14.5, it is clear that the UTS is highly subtle to the changes in welding speed (C). As such, for maximum UTS parameter "C" is the most significant FSW process parameter, followed by shoulder diameter (A) and rotational speed (B). The main effects plots are given in Fig. 14.3. It is found from the plots that the optimal setting of the FSW parameters yielding maximum UTS is $A_3B_2C_3$, that is, A at 18 mm, B at 710 rpm and C at 100 mm/min. The supreme UTS of 142.1 MPa is obtained at optimized parametric combination, which is 15% more compared to BM UTS averaging 122.8 MPa.

ANOVA was used to extricate the relevant conclusion from UTS values. It helps in assessing the significance of each FSW parameter and their interaction on the response. Values of prob > F < 0.05, between 0.05 and 0.1 and above 0.1 indicate that parameters are most significant, significant and insignificant, respectively. From the results of ANOVA (Table 14.5), it is observed that the welding speed (C) is significantly affecting the UTS. The percent (%) contribution was calculated to indicate

Table 14.4 Response table for S/N ratio (UTS)

Level	A	B	C
1	40.66	40.76	40.33
2	41.58	41.65	41.20
3	41.65	41.48	42.36
Delta	0.99	0.89	2.16
Rank	2	3	1

Table 14.5 Analysis of variance for S/N ratios

Source	Degree of freedom	Sum of squares	Mean square	F values	% contribution	*P* value
A	2	1.807	0.903	2.68	18.021	0.27
B	2	1.342	0.671	1.99	13.38	0.33
C	2	6.204	3.10	9.21	61.87	0.098
Residual	2	0.673	0.336		6.71	
Total	8	10.027				

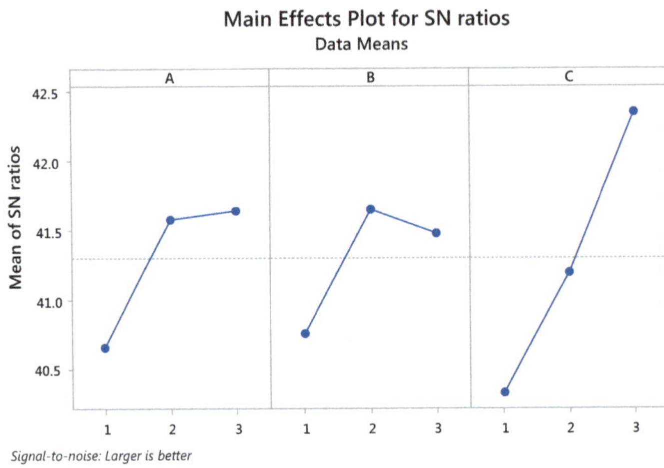

Fig. 14.3 Main effects plot of S/N ratio (UTS)

the order of importance of each parameter and interaction with respect to maximum UTS. It is evident from Table 14.5 that the most dominating factor affecting UTS is welding speed (C), contributing 61.87%, followed by shoulder diameter (A) with 18.02% and lastly, tool rotational speed (B) with 13.38%.

14.3.1.1 Effect of Process Parameters on UTS

It is discernible from Fig. 14.3 that the strength of the joint increases as the shoulder diameter is increased from level 1 to 2. With further increase in tool diameter, the UTS again increases. At low shoulder diameter of 14 mm, insufficient heat and inadequate mixing and material movement might be the reasons for lowest UTS. A larger tool shoulder of 18 mm appears to harvest suitable heat and mixing of the BM without declining the joints UTS. With the increase in tool rotational speed, there is inflation in UTS of the weld up to maximum 710 rpm, followed by diminishing in UTS (Fig. 14.3). When the rotational speed is low (560 rpm), there is insufficient mixing between the plates due to limited temperature generation and insufficient BM deformation. On the other hand, at high rpm (900 rpm), excessive heat generation might have resulted in turbulent material flow and local softening, thereby causing grain coarsening [5, 13]. The joints produced at 710 rpm possess better mechanical properties owing to ample heat generation and material flow required for satisfactory softening and mingling of the deformed BM, leading to refined grains and enhanced UTS [14]. It is clear from the plot that as the weld speed increases, UTS also increases throughout the process. At lower welding speed (63 mm/min), there will be excessive material mixing between the plates. Also, the heat generation and heat transfer from tool to workpiece is higher, resulting in lower UTS. As the welding speed is increased

(80–100 mm/min), the effect of heat generation reduces, leading to more refined grain structure and improved UTS.

14.3.2 Microstructure

The microstructure of the different zones obtained at optimized parameter yielding maximum strength is demonstrated in Fig. 14.4. The three distinct zones, namely stir zone (SZ), thermo-mechanically affected zone (TMAZ) and heat-affected zone (HAZ), are observed during FSW, both in advancing side (AS) and retreating side (RS) (15). The mechanical properties of the joints largely depend on the microstructure. From the micrographs it is observed that there is an appreciable discrepancy in grains in different zones. Dynamically recrystallized equiaxed fine grains are experiential in SZ. Fineness in grains is due to dynamic recrystallization caused by frictional heating and plastic deformation caused by the rubbing and stirring of the tool with the BM [15]. In the TMAZ, grains are not as fine and recrystallized as associated to SZ, due to low heat input and deformation [16]. Highly distorted grains of uneven size can be seen in TMAZ, both in AS and RS. The HAZ involves only welding cycles and no deformation. Non-deformed coarse grains slanting in direction of rolled BM are clearly visible in this zone. Also, the grains are visibly coarser in AS as compared with RS. This may be due to higher temperature experienced at the AS.

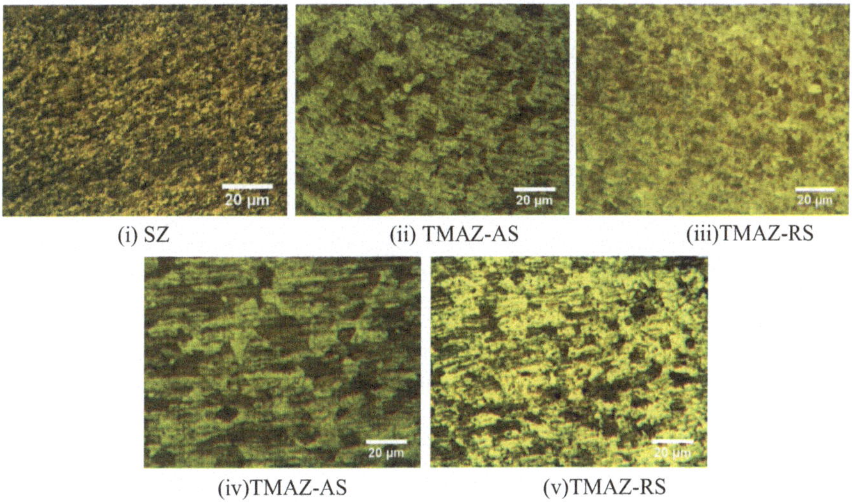

Fig. 14.4 Microstructure evolution for experiment no. 08

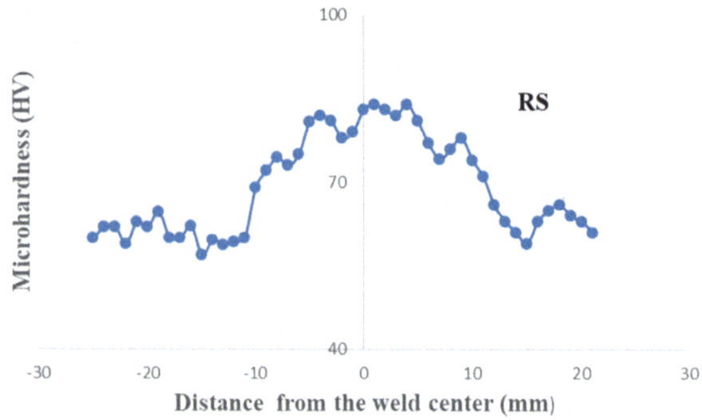

Fig. 14.5 HV profile of the weld produced for experiment no. 8

14.3.3 Microhardness

Microhardness variation along the friction stir welded joint is influenced by the joint integrity and its microstructure at various zones. The microhardness (HV) profile is shown in Fig. 14.5. For these types of AAs, the HV values are generally controlled by size and distribution of grains and solid-solution strengthening [5].

The average HV of the BM was found to be 70 HV. The maximum hardness greater than BM was obtained in the SZ (84 HV to 77 HV). The microhardness in SZ was also greater than the TMAZ/HAZ microhardness. Further HV values were slightly better in RS than the AS. The fine equiaxed grains due to dynamic recrystallization have resulted in the higher HV in SZ. The HV values in TMAZ/HAZ are lower than the SZ due to grain growth. In experiment no. 8 the HV is found to be good due to ample heat generation and material flow required for satisfactory softening and mingling of the deformed BM leading to refined grains.

14.4 Conclusion and Future Scope

FSW of AA 8011 was effectively carried out on 3 mm plates according to Taguchi L9 OA. Some of the noticeable findings are:

1. The best setting of the FSW parameters yielding highest UTS is $A_3B_2C_3$, that is, the shoulder diameter (A) at 18 mm, rotational speed (B) at 710 rpm and welding speed (C) at 100 mm/min.
2. The highest UTS of 142.1 MPa is obtained at optimized parametric combination, which is 15% more as compared to BM.
3. The most dominating factor affecting UTS is welding speed contributing 61.87%, followed by shoulder diameter and tool rotational speed.
4. Dynamically recrystallized equiaxed fine grains were observed in SZ. Highly distorted grains of uneven size were seen in TMAZ. Non-deformed coarse grains were observed in HAZ.

5. The microhardness in SZ was greater than the BM, TMAZ and HAZ microhardness.

The future studies that can be performed pertaining to present work are advocated in the following:

1. The joining of dissimilar materials using FSW method is limited and can be done.
2. Numerous novel and recent computational and analytical models can be used for better understanding of the material flow.
3. Considering the widespread applications of this technique, there is a necessity for appropriate evaluation of the fracture toughness of the FSW joints.
4. The FSW of thick plates can be quite challenging and needs further investigation.

References

1. Cam, G., İpekoğlu, G.: Recent developments in joining of aluminum alloys. Int. J. Adv. Manuf. Technol. **91**(5–8), 1851–1866 (2017)
2. Lan, S., Liu, X., Ni, J.: Microstructural evolution during friction stir welding of dissimilar aluminum alloy to advanced high-strength steel. Int. J. Adv. Manuf. Technol. **82**(9–12), 2183–2193 (2016)
3. Thomas, W.M., Nicholas, E.D., Needham, J.C., Murch, M.G., Temple-Smith, P., Dawes, C.J.: Friction Stir Butt Welding International Patent Application No. PCT/GB92/02203 and Great Britain Patent Application No. 9125978.8. (1991)
4. Sharma, N., Khan, Z.A., Siddiquee, A.N., Shihab, S.K., Wahid, M.A: Effect of process parameters on microstructure and electrical conductivity during FSW of Al-6101 and Pure Copper. Mater. Res. Express **5**, 046519 (2018)
5. Wahid, M.A., Siddiquee, A., Khan, Z., Asjad, M.: Friction stir welds of Al alloy-Cu: an investigation on effect of plunge depth. Arch. Mech. Eng. **63**, 619–634 (2016)
6. Wahid, M.A., Khan, Z.A., Siddiquee, A.N.: Review on underwater friction stir welding: a variant of friction stir welding with great potential of improving joint properties. Trans. Nonferrous Met. Soc. China. **28**(1), 193–219 (2018)
7. Kumar, R., Singh, K., Pandey, S.: Process forces and heat input as function of process parameters in AA5083 friction stir welds. Trans. Nonferrous Met. Soc. China **22**, 288–298 (2012)
8. Wahid, M.A., Khan, Z.A., Siddiquee, A.N., Shandley, R., Sharma, N.: Analysis of process parameters effects on underwater friction stir welding of AA 6082-T6. Proc. Inst. Mech. Eng. Part B J. Eng. Manuf. **233**(6), 1700–1710 (2019). https://doi.org/10.1177/0954405418789982
9. Lakshminarayanan, A.K., Balasubramanian, V.: Process parameters optimization for friction stir welding of RDE-40 aluminium alloy using Taguchi technique. Trans. Nonferrous Metals Soc. China **18**(3), 548–554 (2008)
10. Jayaraman, M., Sivasubramanian, R., Balasubramanian, V., Lakshminarayanan, A.K.: Optimization of process parameters for friction stir welding of cast aluminium alloy A319 by Taguchi method. J. Sci. Ind. Res. **68**, 36–43 (2009)
11. Wahid, M., Siddiquee, A.N., Khan, Z.A., Sharma, N.: Analysis of cooling media effects on microstructure and mechanical properties during FSW/UFSW of AA 6082-T6. Mater. Res. Express **5**(4) 2018
12. Rajakumar, S., Balasubramanian, V.: Establishing relationships between mechanical properties of aluminium alloys and optimised friction stir welding process parameters. Mater. Des. **40**, 17–35 (2012)
13. Zhang, H., Liu, H.: Mathematical model and optimization for underwater friction stir welding of a heat-treatable aluminum alloy. Mater. Des. **45**, 206–211 (2013)

14. Wang, Q., Zhao, Y., Yan, K., Lu, S.: Corrosion behavior of spray formed 7055 aluminum alloy joint welded by underwater friction stir welding. Mater. Des. **68**, 97–103 (2015)
15. Çevik, B., Özçatalbas, Y., Gülenç, B.: Friction stir welding of 7075-T651 aluminium alloy. Pract. Metallogr. **53**(1), 6–23 (2016)
16. Çevik, B., Özçatalbas, Y., Gülenç, B.: Effect of welding speed on the mechanical properties and weld defects of 7075 Al alloy joined by FSW. Kovove Mater. Met. Mater. **54**(4), 241–247 (2016)

Chapter 15
Study the Effects of Input Process Parameters in V-Bending Die on Springback/Go for Different Aluminium Alloys Using Taguchi Approach

Praveen Kumar and Satpal Sharma

Abstract Springback effect is the most sensitive feature in sheet metal forming. There are many input process parameters of die and press, which affect the springback/go. In the present study, three different aluminium alloys (AA), namely AA6082-T6, 5083-O and 5052-H32, are used. The objective of this work is to compare the effect of process parameters of V-bending die on springback by using Taguchi method. Bending angle (BA), punch load (PL) and punch velocity (PV) were taken as input process parameters. Taguchi approach and ANOVA (analysis of variance) were applied to find the most significant parameter(s) which affects the output response, that is, springback/go in V-bending process. The experiments were performed on V-bending die using a blank size $40 \times 100 \times 2.0$ mm^3 and the springback/go value was calculated using optical microscope. Taguchi L_9 orthogonal array design of experiment was applied to all aluminium alloys AA6082-T6, AA5052-H32 and AA5083-O and the experiments were performed separately. The result shows that the bending angle is the most significant parameter affecting the springback/go value with a contribution of approximate 99% for all aluminium alloys. The optimal values obtained for springback/go for AA6082-T6 at 120° BA, 4 KN PL and 8 mm/s PV, for AA5083-O at 120° BA, 2 KN PL and 24 mm/s PV and for AA5052-H32 are 120° BA, 4 KN PL and 8 mm/s PV.

Keywords V-bending die · CAD model · Springback · Spring go · Taguchi method · ANOVA

P. Kumar (✉) · S. Sharma
Gautam Buddha University, Gautam Buddha Nagar 201312, UP, India
e-mail: praveen.jbm@gmail.com

P. Kumar
Delhi Institute of Tool Engineering, Okhla, New Delhi 110020, India

© Springer Nature Singapore Pte Ltd. 2020

I. Singh et al. (eds.), *Trends in Manufacturing Processes*,
Lecture Notes on Multidisciplinary Industrial Engineering,
https://doi.org/10.1007/978-981-32-9099-0_15

15.1 Introduction

5xxx and 6xxx series are different types of aluminium alloys that have wide range of applications in the automobile, defence and aerospace industries. AA6082, AA5083 and AA5052 are commonly used for these industries. In V-bending process, the tension applied to the sheet material must be between the elastic limit and the maximum tensile stress. When the load is removed, springback or spring-go occurs according to the tension structure and the magnitude of the tension within the material. The different parameters of V-bending die and press, that is, die radius, width of the blank, thickness variation, press force, holding time and so on affects the springback and spring-go. Thipprakamas and Rojananan [1] investigated the punch height effect on springback/go phenomenon in the V-bending process. With respect to the angular punch radius, the FEM simulation results showed that the springback/go phenomenon could be theoretically clarified based on material flow analysis and the stress distribution analysis. The amount of spring-go decreased as the angular punch radius increased and the amount of springback decreased as the angular punch radius decreased. Hakan et al. [2] investigated the effects of the material properties and punch tip radius of the spring-forward in V-bending process. It was observed that the spring-forward was observed due to the material properties and 90° V-bending process. Thipprakmas [3] investigated the effect of punch height on V-bending using FEM. FEM was used to investigate the effects of bending angle in the V-bending process. With respect to the punch heights, FEM simulation results showed that the effects of punch height of the bending angle could be theoretically clarified based on the material flow analysis and the stress distribution analysis. Özdemir [4] determined the microstructure changes during deformation and concluded that 60° and 90° bending angles showed parallelism. They observed an inverse proportion with the results of 30° bending angles. The greatest amount of spring-go was observed in the normalized materials and the lowest amount of spring-go was observed in the tempered sheet materials. Özdin et al. [5] investigated the effect of springback/spring-go of AISI 400S and observed that as the bending angle decreases, spring-go increases and as the punch radius is increased, spring-go increases. Jooybari et al. [6] studied the effects of significant parameters, including sheet thickness, sheet anisotropy and punch tip radius, on springback/spring-go into V-die and U-die bending processes of CK67 (DIN 17222) steel sheet have been conducted. The bending angle to the rolling direction influences the springback and the spring-go, the greater is the angle to the rolling direction. Thipprakmas et al. [7] examined the effects of process parameters, such as bending angles, sheet thickness and stress distributions. It was determined that the bending angles and sheet thickness had a great effect on bending processes for both springback and spring-go and observed that Taguchi method and ANOVA technique were effective tools to predict the degree of importance of process parameters in the V-bending process. Chikalthankar et al. [8] reviewed and examined the factors affecting springback in sheet metal. Springback in sheet metal bending depends on different variable parameters, like punch angle, sheet thickness, punch tip radius, punch height, bending force, clearance between punch and die, sheet anisotropy. It was observed that sheet thickness is an important variable parameter affecting the springback and residual stresses. Abdullah and Samad [9] experimentally investigated springback of AA6061 aluminium alloy strip via V-bending process. The effect of length, thickness and bend angle has been studied and the result depicted that the springback was significant and related to the thickness and bend angle changes. Gautam et al. [10] studied the analytical effect of thickness variation of CRDQ sheet on springback in V-bending die. Hyperform software was used with the thickness variation of 2.0–4.0 mm. It was found that the value of springback value decreases with increase of sheet thickness value. Datta et al. [11] carried a study experimentally and analytically on the effect of process parameter on springback in aluminium alloy sheet AA6061-T6 using response surface methodology. It was observed that the decrease in bending angle and increase in the blank width size and punch velocity, the springback increases. Adnan et al. [12] investigated the springback effect on aluminium alloy AA6061 with non-uniform thickness section using Taguchi method. It was observed that the bend angle strongly influences the springback and the more contributions of the bend radius to the percentage of springback are 92.51%. Gautam and Kumar [13] examined the springback effect in V-bending die of tailor-welded blank of interstitial-free steel. It was observed that the weld zone properties affect the springback than the anisotropic of sheet. Liu et al. [14] investigated on the bending accuracy in micro W-bending using Taguchi L_8 orthogonal array. It was found that the predicted value fits well with the experimental results. The foil thickness contributes the most, 79.03% over foil orientation and grain size. Gedekar et al. [15] optimized the process parameter affecting the springback for CR2 grade steel sheet of IS513-2008 by using Taguchi method. It was found that the punch angle has more contribution with 62.92% effect on springback and finally concluded that as the punch angle and the die opening increases, the springback effect reduces. The present work comprises the experimental comparative study on the effect of process parameter on springback or spring-go in V-bending die. Three different aluminium alloys AA6082-T6, AA5083-O and AA5052-H32 were used for this work. Taguchi method was used to predict the importance of process parameters for the given material.

15.2 Research Methodology

15.2.1 Materials and Method

Three different types of aluminium alloys AA6082-T6, AA5052-H32 and AA5083-O were used for this study. These aluminium alloys are widely used in bending, forming and drawing operations, when there is requirement of lightweight materials. Blank size of $40 \times 100 \times 2.0\ mm^3$ was prepared from the stock material. The blank sizes were cut by shearing machine and the surface was properly finished. Burr free edges were obtained and used for performing the experiments in V-bending die. The tensile specimen of standard size for all material as per the ASTM E8M-04 standard has been prepared by water jet machining process. Samples were made to determine the mechanical properties using universal testing machine (UTM) given in Table 15.1. The chemical composition of all materials was checked by the energy dispersive X-ray analyser. The percentages of constituent materials in the alloy are given in Table 15.2.

15.2.2 Design of Experiment and Process Parameters Used for Study

The objective of Taguchi method is to produce a good quality of the product at low cost. Taguchi method created a small number of experiments to study all parameters that affect the output response. The important significant factor was considered as controlling factor. According to the Taguchi method, L_9 orthogonal array was designed for the experiment. Totally, nine experiments are to be conducted for all the aluminium alloys. The input parameters such as bending angle, punch load and punch velocity were chosen for the experiment based on the literature study shown in Table 15.3 [11]. As per Taguchi design method, the number of experiments was defined. MINITAB 17.0 software was used to create the three-level designs and L_9 orthogonal array for selected levels of input process parameter, which is shown in Table 15.4. Taguchi experiment was applied to all aluminium alloys AA6082-T6, AA5052-H32 and AA5083-O and performed separately. The value of springback/go angle was taken as output response. The results were transformed into a signal-to-noise (S/N) ratio to measure quality characteristics. The characteristic "smaller is better" for signal-to-noise ratio was considered.

15.2.3 Fabrication of V-Bending Die and Trial on a Hydraulic Press

As per the V-shaped component, bending die was designed on CAD software [11]. Three different die sizes and punch sizes of 30°, 45° and 120° were used and manufactured by the CNC milling machine. The complete V-bending die consists of back post die set (lower plate, upper plate, guide pillar and guide bush), punch and die has been fabricated and assembled with the help of screws and dowels. The material of die and punch was taken D2 steel with HRC 55–58. V-bending die was used to perform for all the aluminium alloys. As per the calculation of the bending force, 1.5 KN maximum bending force was required for V-bending die for bent component, as shown in Fig. 15.1. The total 2 KN force was required for the hydraulic press. Hydraulic press, that is, the single action press having capacity of 100 KN was used for the trial samples. V-bending die was successfully installed in the press in which clamped upper half of the die (upper plate and punch) with press Ram and lower

Table 15.1 Chemical properties of (1) AA6082-T6; (2) 5083-O; (3) 5052-H32

No.	Material name	Manganese (Mn)	Iron (Fe)	Magnesium (Mg)	Silicon (Si)	Copper (Cu)	Zinc (Zn)	Titanium (Ti)	Chromium (Cr)	Aluminium (Al)
1	6082-T6	0.83	0.60	1.33	0.30	0.10	0.20	0.10	0.37	95.54
2	5083-O	0.83	0.48	4.37	0.10	0.10	0.44	0.23	0.05	93.4
3	5052-O	0.1	0.48	2.72	0.25	0.45	0.1	0.30	0.32	95.28

Table 15.2 Mechanical properties of (1) AA6082-T6; (2) 5083-O; (3) 5052-H32

No.	Material name	Properties						
		Density (g/cc)	Young's modulus (GPa)	Yield strength (MPa)	Tensile strength (MPa)	Elongation (%)	Shear strength (MPa)	Hardness vickers (HB)
1	AA6082-T6	2.7	70	262	301	13	210	110
2	AA5083-O	2.8	73	143	241	24	180	100
3	AA5052-H32	2.7	70	125	170	10	124	70

Table 15.3 Parameters selected and their levels

Input parameter	Levels			
	Units	−1	0	1
Bending angle	Degrees	30	45	120
Punch load	KN	2	4	6
Punch velocity	mm/s	8	16	24

Table 15.4 Design of experiment L_9 orthogonal array for selected levels of critical parameters

Run no.	Bending angle (°)	Punch load (KN)	Punch velocity (mm/s)
1	30	2	8
2	30	4	16
3	30	6	24
4	45	2	16
5	45	4	24
6	45	6	8
7	120	2	24
8	120	4	8
9	120	6	16

(a) (b) (c) (d)

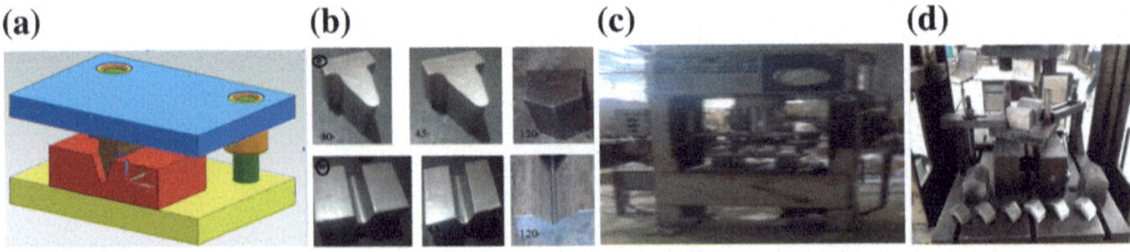

Fig. 15.1 **a** CAD drawing. **b** Bending punch and die. **c** Hydraulic press. **d** Bent samples

half clamped to lower bolster plate of the press with the help of riser plates and strap clamp. First, check the operation and alignment of die before start of the experiment. According to the design of experiments, nine experiments were performed for each aluminium alloy AA6082-T6, 5083-O and 5052-H32. Each experiment was repeated two times for better result.

15.3 Results and Discussion

Experiments have been performed successfully and the V-bend component was checked by the optical microscope in degree level. A total of nine experiments were performed for all the aluminium alloys AA6082-T6, AA5083-O and AA5052-H32. The springback/go values of all different aluminium alloys were shown in Table 15.5. Taguchi approach was used to perform the design of experiments. The experimental data analyzed the signal-to-noise ratio (S/N ratio). The S/N ratio was used to find out the optimal process parameter from each level. The response table for means and S/N ratio was calculated. The rank in a response table was used to identify which input parameters have the significant effects. The delta value of factor is high shows the rank number is 1. The main effect plot for S/N ratio and the means was drawn for the AA6082-T6, AA5083-O and AA 5052-H32. The graph between input parameters and S/N ratio for both the cases. Here, the input parameter represents the signal where the parameter was controllable by the users and noise cannot be controllable by the users. S/N ratio affects the springback values due to external factors. In this experimental study, the outcome is springback/go, so lesser the springback value, there will be better quality. The Taguchi method was discussed separately for all aluminium alloys AA6082-T6, AA5083-O and AA5052-H32.

Table 15.5 Springback value for all aluminium alloys

Run no.	Bending angle (°)	Punch load (KN)	Punch velocity (mm/s)	Springback/go (°)		
				AA6082-T6	AA5083-O	AA5052-H32
1	30	2	8	37.36	36.12	32.13
2	30	4	16	36.13	34.58	32.31
3	30	6	24	32.21	32.39	31.8
4	45	2	16	53.35	48.23	49.18
5	45	4	24	52.32	46.56	47.45
6	45	6	8	48.3	45.23	46.8
7	120	2	24	121.1	120.2	120.5
8	120	4	8	120	119.58	119.8
9	120	6	16	118.8	118.56	118.8

15.3.1 For AA6082-T6

The main effect plot for means and S/N ratio is shown in Fig. 15.2. These graphs represent the effect of factors and its values on the output response, that is, springback/go. The reference line was drawn and shows the average S/N ratio for all input factors, that is, bending angle, punch load and punch velocity. From the main effect plot, for S/N ratio, it has been clearly shown that the springback will be less at 120° bending angle, 2 KN punch load and 16 mm/s of the punch velocity. In Table 15.6, delta value of the means is indicated. The value 35.23 indicates the means of the mean value of springback to the first level of parameter, that is, bending angle on 30°. Similarly, all the mean values of springback for all levels of input parameters were calculated. The delta value 84.73 is maximum for the bending angle as input parameter other than remaining two parameters, that is, punch load and punch velocity. So, the value is given 1 for bending angle and most affecting parameter for the springback in V-bending process for AA6082-T6. In Table 15.7, the delta value gives the variation in the S/N ratio within the level. The value of delta shows that every factor was contributed in a response according to the response table for S/N ratio. The bending angle has more affecting parameter within delta value of 10.66 when compared to punch load and punch velocity.

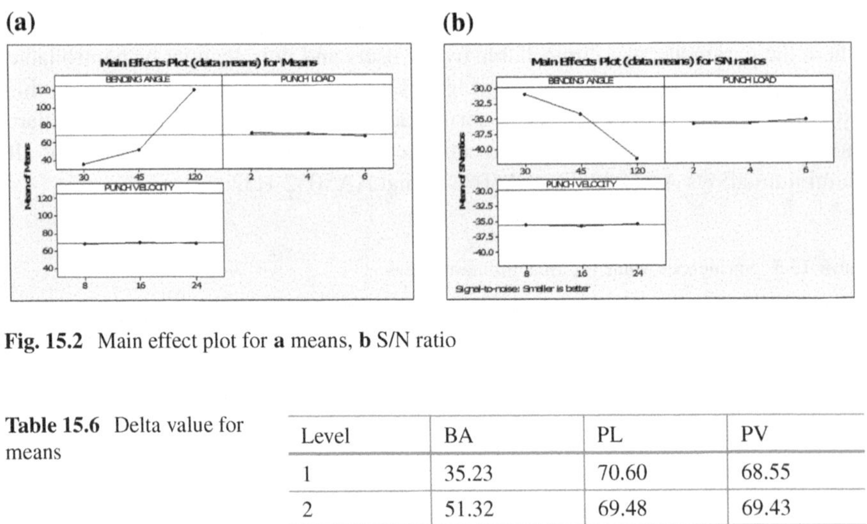

Fig. 15.2 Main effect plot for **a** means, **b** S/N ratio

Table 15.6 Delta value for means

Level	BA	PL	PV
1	35.23	70.60	68.55
2	51.32	69.48	69.43
3	119.97	66.44	68.54
Delta	84.73	4.17	0.88
Rank	1	2	3

Table 15.7 Response table for S/N ratios at "smaller is better"

Level	BA	PL	PV
1	−30.92	−35.88	−35.57
2	−34.20	−35.70	−35.73
3	−41.58	−35.11	−35.40
Delta	10.66	0.77	0.33
Rank	1	2	3

15.3.2 For AA5083-O

Similar to AA6082-T6, nine experiments were performed as per the design of experiments by using the Taguchi approach. The main effect plot for means value and S/N ratio and graph between input factor and springback values are shown in Fig. 15.3. Similarly, the main effect plot for S/N ratio shows that the springback/go value will be less at 120° bending angle, 2 KN punch load and 8 mm/s of punch velocity, as shown in Fig. 15.3. The mean value of a springback for the first level of parameter, that is, bending angle is 30°. Similarly, in second and third level, the values of the mean were 46.67 and 119.45, respectively. Similarly, all the mean values of springback at all levels of factors were calculated. The delta value (85.08) was high for bending angle, as shown in Table 15.8. The delta value for bending angle was greater than the other two input parameters, that is, punch load and punch velocity. In Table 15.9, the response table for S/N ratio at the characteristic "smaller the better" was taken.

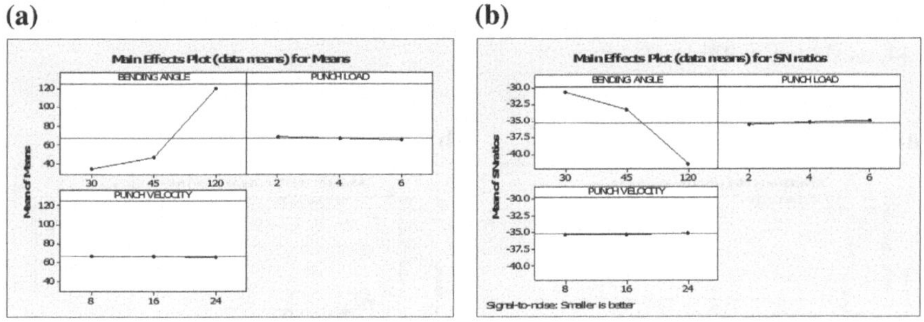

Fig. 15.3 Main effect plot for **a** means, **b** S/N ratio

Table 15.8 Delta value for means

Level	BA	PL	PV
1	34.36	68.18	66.98
2	46.67	66.91	67.12
3	119.45	65.39	66.38
Delta	85.08	65.39	66.38
Rank	1	2	3

Table 15.9 Response table for S/N ratios at "smaller is better"

Level	BA	PL	PV
1	−30.71	−35.47	−35.27
2	−33.38	−35.23	−35.31
3	−41.54	−34.93	−35.06
Delta	10.83	0.54	0.25
Rank	1	2	3

The delta value was higher at bending angle, 10.83, in comparison to punch load of 0.54 and punch velocity of 0.25, which means this factor affects the springback/go value more and its contribution in a particular response is more.

15.3.3 For AA5052-H32

Similarly, for the AA5052-H32, the same procedure was applied to analyse the data of experimental result using Taguchi design of experiment. The main effect plot for means and the S/N ratio graph was generated and shown in Fig. 15.4. It was clearly shown that the springback/go value will be less at 120° bending angle, 2 KN punch load and 16 mm/s of punch velocity. Delta value was calculated and found that the highest value is of bending angle, that is, 87.62 when compared with other two parameters, punch load with a value of 1.6 and punch velocity with a value of 0.52, as shown in Table 15.10. The delta value of S/N ratio was higher at bending angle, that is, 11.44 in comparison to punch load delta value of 0.21 and punch velocity 0.44, as shown in Table 15.11.

(a) **(b)**

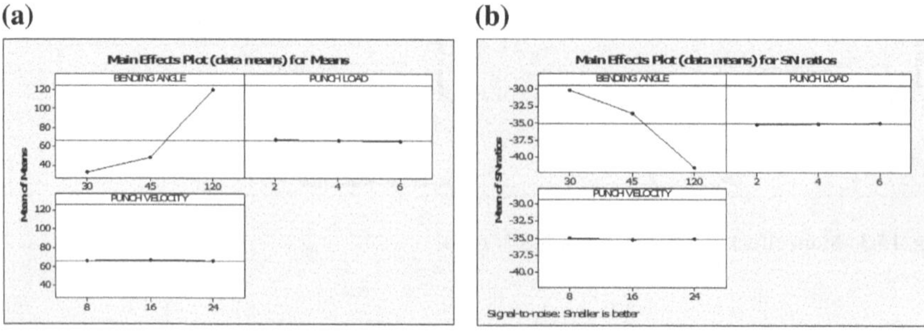

Fig. 15.4 Main effect plot for **a** means, **b** S/N ratio

Table 15.10 Delta value for means

Level	BA	PL	PV
1	32.08	67.27	66.24
2	47.81	66.27	66.76
3	119.70	65.80	66.58
Delta	87.62	1.47	0.52
Rank	1	2	3

Table 15.11 Response table for S/N ratios at "smaller is better"

Level	BA	PL	PV
1	−30.12	−35.20	−35.04
2	−33.59	−35.09	−35.17
3	−41.56	−34.98	−35.06
Delta	11.44	0.21	0.14
Rank	1	2	3

15.4 ANOVA (Analysis of Variance) Table and Regression Equation for Response

ANOVA is a standard statistical method to test difference between mean values. It was carried out to find the influence of parameters among the input parameters. ANOVA table was made for all design of experiments for all the materials separately. Finally, it was considered that the bending angle has highest value of contribution than other parameters. It was shown that the 99% contribution of bending angle has major influence. Punch load and punch velocity have shown very less contribution with 0.2% effect on springback. The experimentation result with % contribution of relative parameter of ANOVA table for AA6082-T6, AA5052-H32, AA 5083-O is shown in Table 15.12.

As per the Taguchi approach, the optimum level of parameters was achieved where the S/N ratio has more value. For calculation, the value of error for the said formula was used. % error = {(prediction value − actual value)/prediction value} × 100. The percentage of error between the predicted value and confirmation test result was examined and shown in Table 15.13.

A regression equation was measured to calculate and predict the springback/spring-go value for given aluminium alloys.

For AA6082-T6

$$\text{Springback/Go (degree)} = 12.4 + 0.933\ \text{BA} - 1.04\ \text{PL} - 0.0006\ \text{PV} \quad (15.1)$$

For AA5083-O

$$\text{Springback/Go (degree)} = 8.24 + 0.953\ \text{BA} - 0.698\ \text{PL} - 0.0371\ \text{PV} \quad (15.2)$$

Table 15.12 Experimentation result of ANOVA table

Source	DF	Adj SS	Adj MS	F	Contribution (%)
For 6082-T6					
Bending Angle	2	12228.2	6114.1	4575.71	99.80
Press Load	2	19.3	9.6	7.21	0.15
Punch Velocity	2	2.3	1.2	0.87	0.02
Error	2	2.7	1.3		0.02
Total	8	12252.5			
For 5083-O					
Bending Angle	2	12686.7	6343.3	43404.55	99.80
Press Load	2	11.7	5.9	40.04	0.1
Punch Velocity	2	0.9	0.5	3.15	0.007
Error	2	0.3	0.1		0.002
Total	8	12699.6			
For 5052-H32					
Bending Angle	2	13092.9	6546.4	13635.56	99.9
Press Load	2	3.2	1.6	3.38	0.02
Punch Velocity	2	0.4	0.2	0.44	0.003
Error	2	1.0	0.5		0.007
Total	8	13097.5			

Table 15.13 Optimal value for springback/go

Material	BA	PL	PV	% error
AA6082-T6	120	4	8	0.2
AA-5083-O	120	2	24	0.13
AA5052-O	120	4	8	0.33

For AA5052-H32

$$\text{Springback/Go (degree)} = 4.69 + 0.969\ \text{BA} - 0.368\ \text{PL} + 0.0213\ \text{PV} \quad (15.3)$$

15.5 Conclusions

In this present study, three input process parameters, bending angle, punch load and punch velocity, were used. Taguchi method along with ANOVA technique was used for optimization of these process parameters. Three different aluminium alloys AA6082-T6, AA5083-O and AA5052-H32 were used, and the same experiments were performed. According to the results, the comparative conclusions on the effect of process parameter for aluminium alloys are as follows:

1. In all the cases, the springback/go value is less at the bending angle of punch 120°. The final bending angles are 120° for AA6082-T6, 118.56° for AA5083-O and 118.8° for AA5052-H32.
2. In all the cases, the bending angle was significant parameter affecting the springback/go with the contribution of approximate 99% in comparison with other two parameters.
3. Optimal values lie for springback/go for the material AA6082-T6 at 120° BA, 4 KN PL and 8 mm/s PV, for the material AA5083-O at 120° BA, 2 KN PL, 24 mm/s and for the material AA5052-H32 at 120° BA, 4 KN PL and 8 mm/s PV.
4. In all the cases, the punch load increases from 2 to 6 KN, the bending force increases and then the springback decreases.
5. In all the cases, the punch velocity increases from 8 to 24 mm/s, then the springback decreases.
6. In all the cases, the springback value was found at bending angles 30° and 45° only, but the spring-go value was found at 120° bending angle and that the value range reaches to 118–120°.
7. Regression equation was measured to predict the spring back/go value for all the materials.

References

1. Thipprakmas, S., Rojananan, S.: Investigation of spring-go phenomenon using finite element method. Mater. Des. **29**, 1526–1532 (2008)
2. Hakan, D., Mustafa, O., Murat, S.: Effects of material properties and punch tip radius on spring-forward in 90° V bending processes. J. Iron Steel Res. Int. **20**(10), 64–69 (2013)
3. Thipprakmas, S.: Finite element analysis of punch height effect on V-bending angle. Mater. Des. **31**, 1593–1598 (2010)
4. Özdemir, M.: Characterization of microstructure and bending response of sheet material: influence of thickness. J. Adv. Mater. Process. **3**, 3–13 (2015)
5. Özdin, K., Büyük, E., Abdalov, F., Bayram, H.: Investigation of spring-back and spring-go of AISI 400 S sheet metal in “V” bending dies depending on bending angle and punch radius. Appl. Mech. Mater. **532**, 549–553 (2014)
6. Bakhshi-Jooybari, M., Rahmani, B., Daeezadeh, V., Gorji, A.: The study of spring-back of CK67 steel sheet in V-die and U-die bending processes. Mater. Des. **30**, 2410–2419 (2009)
7. Phanitwong, W., Sontamino, A., Thipprakmas, S.: Effects of part geometry on spring-back/spring-go feature in U-bending process. Key Eng. Mater. **549**, 100–107 (2013)
8. Chikalthankar, S.B., Belurkar, G.D., Nandedkar, V.M.: Factors affecting on springback in sheet metal bending : a review. IJEAT 247–251 (2014)
9. Abdullah, A.B., Samad, Z.: An experimental investigation of springback of AA6061 aluminum alloy strip via V-bending process. IOP Conf. Ser. Mater. Sci. Eng. **50**, 012069 (2013)
10. Chandra, V., Kumar P., Gautam, B., Rawat, K.: Analysis of springback variation in V bending. Int. J. Eng. Res. Technol. **5**, 555–560 (2016)

11. Datta, S., Kumar, P., Chandra, V., Aggarwal, N.: Analytical and experimental evaluation of factors effecting springback in aluminium alloy sheet. 754–758 (2016)
12. Adnan, M.F., Abdullah, A.B., Samad, Z.: Springback behavior of AA6061 with non-uniform thickness section using Taguchi method. Int. J. Adv. Manuf. Technol. (2016). https://doi.org/10.1007/s00170-016-9221-0
13. Gautam, V., Kumar, D.R.: Experimental and numerical investigations on springback in V-bending of tailor-welded blanks of interstitial free steel. Proc. Inst. Mech. Eng. Part B J. Eng. Manuf. (2017). https://doi.org/10.1177/0954405416687146
14. Liu, X., Zhao, S., Qin, Y., Zhao, J.: A parametric study on the bending accuracy in micro W-bending using Taguchi method. Measurement (2016). https://doi.org/10.1016/j.measurement.2016.12.007
15. Gedekar, R.D., Kulkarni, S.R., Kavad, M.B.: Optimization of input process parameters affecting on springback effect in sheet metal 'V' bending process for CR2 grade steel sheet of IS 513-2008 material by using Taguchi method (2018)

Chapter 16
Comparative Study of Particulate Erosion Phenomena Between Elbow and Target Tee in Pipe Flow

Ajay Singh Jamwal

Abstract In offshore oil and gas operations, sand particles with the crude oil causes drastic particulate erosion of transporting pipes, especially during bending of flow, which results in higher economic losses in terms of replacement of pipe fittings. Sand particles having critical velocity strike the wall surface of pipe-elbow, which causes pitting phenomena of material from the erosion zone on the basis of brittleness or ductile nature of pipe fitting material. This problem is countered by using target tee or cushion tee, which is generally a 90° elbow with a blind-end branch. The blind branch is filled with soft materials to accommodate the excessive erosion caused by sand particles. The blind end of target tee is filled with soft material, which provides the cushion effect to the impinging solid particles of fluid at the bend surface. This cushion effect reduces the probability of erosion in target tee and prevents the flow process from the economic losses caused by erosion.

Keywords Target tee/cushion tee · Particulate erosion · Pipe elbow

16.1 Introduction

16.1.1 Mechanism of Erosion in Pipe Flow

Erosion is a phenomenon of pitting of material or removal of material in the form of small chips due to repeated collision of solid particles on the surface material. Various mechanism of erosion takes place in pipe flow. Some of them are as follows:

(i) Cavitation
(ii) Particulate erosion
(iii) Two-phase flow erosion

When water pressure in the pipe reaches vapour pressure then formation of bubbles takes place and when these bubbles reaches an area of high pressure zone, then

A. S. Jamwal (✉)
Guru Nanak Dev Engineering College (GNDEC), Ludhiana, Punjab, India
e-mail: ajaysingh1@rediffmail.com

© Springer Nature Singapore Pte Ltd. 2020

I. Singh et al. (eds.), *Trends in Manufacturing Processes*,
Lecture Notes on Multidisciplinary Industrial Engineering,
https://doi.org/10.1007/978-981-32-9099-0_16

bubbles condenses into liquid. Thus cavity is formed and the surrounding liquid rush to fill in the vacant places and a violent sound is produced. This is called cavitation phenomena. However, in two-phase flow erosion, pitting of material takes place due to variation in velocity of gas and liquid in the fluid mixture.

Of the above all mechanisms, particulate erosion is the most critical erosion phenomenon that takes place during pipe flow, which causes maximum erosion of pipe material.

Particulate Erosion

It is also called sand particle erosion. During the flow, sand particles having critical velocity strike the surface of pipe. The kinetic energy of sand particle is transferred to the pipe surface, which causes removal of material in the form of small chips. Particulate erosion depends on various factors, like pipe material, sand particle size and flow characteristics of fluid.

Brittle and Ductile Erosion

In pipe flow, particulate erosion is a mechanical process of metal removal, which is further divided into two heads on the basis of pipe material characteristics (Fig. 16.1).

If the material of pipe surface is ductile, then due to continuous striking of solid particles at a particular angle on the surface of pipe, it creates a crater-like structure

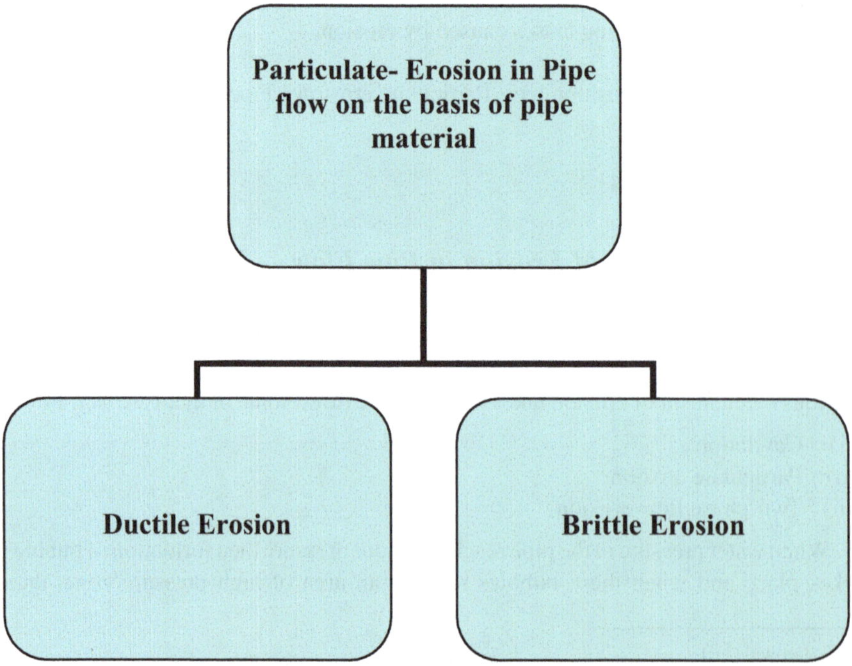

Fig. 16.1 Classification of particulate erosion in pipe flow

which further grows due to continuous impact of solid particles on pipe surface. At the end, it leads to micro-machining of material surface in the form of small chips. Figure 16.2 shows the various stages, which occurred during the ductile erosion phenomena of pipe material.

However, if the material surface of pipe is brittle in nature then it leads to the formation of lateral cracks inside the pipe material surface, when solid particles strike the pipe surface. This further leads to the metal removal process in the form of small pieces of material. Figure 16.3 shows the various steps which take place in brittle erosion of pipe surface.

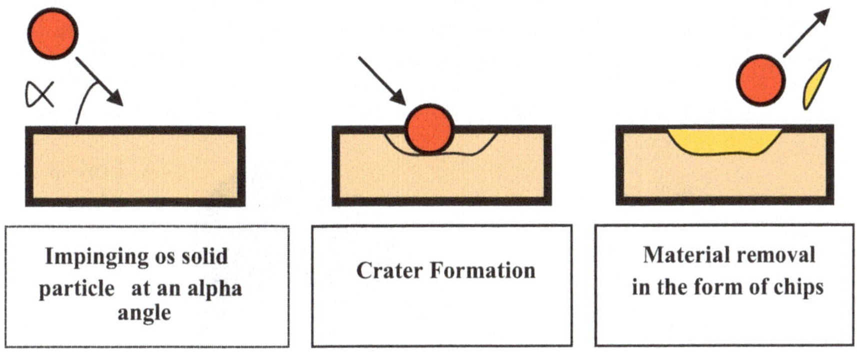

Fig. 16.2 Ductile erosion phenomena in pipe flow

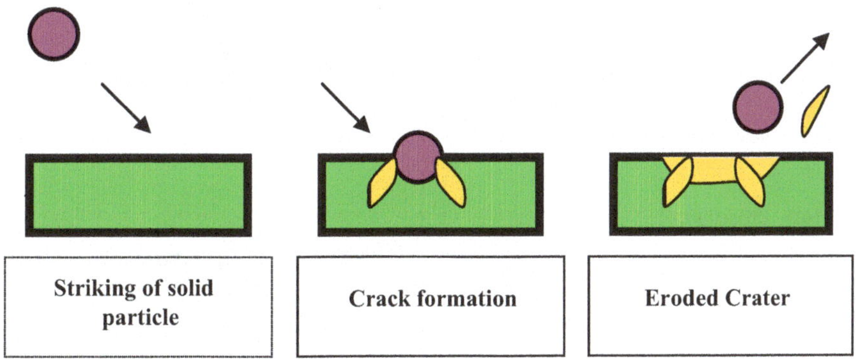

Fig. 16.3 Brittle erosion phenomena in pipe flow

16.2 Erosion Phenomena in Pipe Bend

16.2.1 Mechanism of Erosion in Pipe Flow

Figure 16.4 shows the erosion phenomena that take place in the pipe elbow during the bending of flow regime.

In the flow regime of pipe, solid particles of fluid enter the elbow and strike the bend surface at a critical angle with certain velocity. Owing to its striking, the kinetic energy is transferred to bend surface, which causes erosion of wall thickness at the bend of the pipe. Hence during the pipe flow, maximum erosion takes place at the bend of the pipe, that is, elbow due to sudden change of direction of flow. Figure 16.5 shows the pictorial view of eroded pipe elbow.

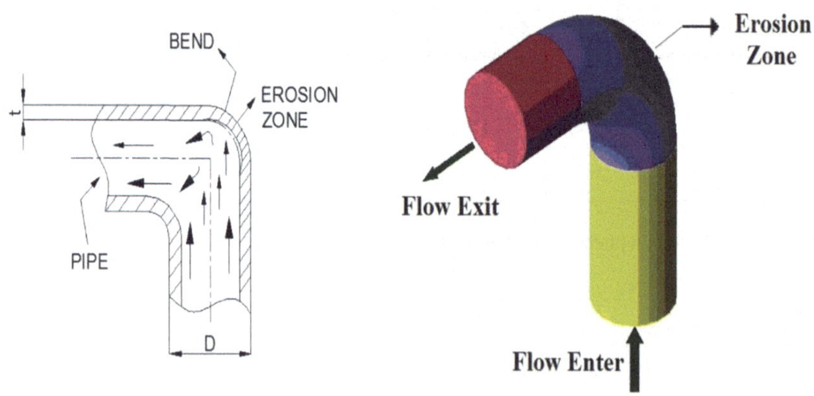

Fig. 16.4 Erosion in pipe elbow

Fig. 16.5 Eroded pipe elbow

Calculation of minimum wall thickness of pipe bends

According to ASME 31.3, the minimum thickness of pipe bend can be determined by the following formulae:

$$t = t_1 + A$$

$$t_1 = \frac{PD}{2[(SQJ/I) + PC]}$$

where t is the minimum thickness of pipe bend after bending; P is the internal design gauge pressure; D is the outside diameter of pipe bend; S is the stress value for material; Q is the quality factor; J is the weld joint strength reduction factor; I is the intrados or extrados factor; C is the value of coefficient; and A is the sum of mechanical allowances plus corrosion and erosion allowances.

16.3 Applicability of Cushion/Target Tee

16.3.1 Mechanism of Erosion in Target Tee

To counter the excessive particulate erosion in pipe fittings like elbow, there is a general practice to use cushion or target tees during sand–fluid flow. Cushion tees are also called as target tees, which are actually a 90° elbow with one branch as blind end (Fig. 16.6).

In cushion tees, the solid particles of fluid enter from one end and strike the bend surface at a critical angle, which causes erosion of surface material. However, the blind end of target tee is filled with soft materials like lead, which reduces the velocity of solid particle impinging over the bend surface. Thus, it will prevent the

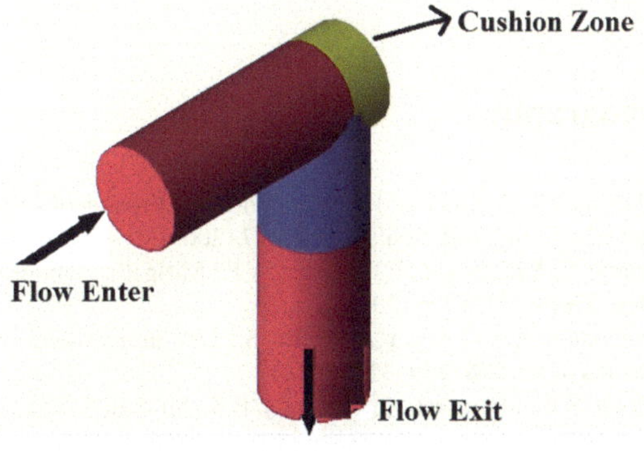

Fig. 16.6 Target tee with cushion zone

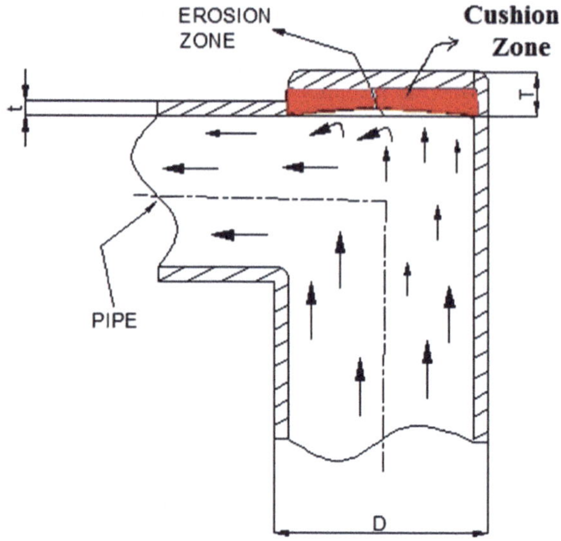

Fig. 16.7 Erosion mechanism in target tee

excessive erosion of wall thickness of pipe bend. The soft material filled at the blind end provides the cushion effect to the impinging solid particles of fluid. Hence target tees are also called as cushion tees (Fig. 16.7).

16.4 Conclusion

In this research work, a comparative study of particulate erosion between elbow and target tee is carried out and it has been found that target or cushion tee suffers less erosion in comparison to elbow during sand–fluid flow in pipe. It has also been revealed that due to the cushion effect provided by the soft material at the blind end branch, the target tee experiences less erosion and thus prevents the flow process from economic losses such as frequent replacement of pipe fittings.

Bibliography

1. Blanchard, D.J., Griffith, P., Rabinowicz, E.: Erosion of a pipe bend by solid particles entrained in water. J. Eng. Ind. **106**(3), 213–217 (2009)
2. Duarte, C.A.R., Souza, F.J., Santos, V.F.: Mitigating elbow erosion with a vortex chamber. C.R. Mec, **336**(9), 731–736 (2008)
3. Lachouette, D., Golay, F., Bonelli, S.: One dimensional modeling of piping flow erosion. J. Powder Tech. **288**, 229–353 (2016)
4. Vieira, R.E., Mansouri, A., McLaury, B.S., Shirazi, S.A.: Experimental and computational study of erosion in elbows due to sand particles in air flow. J. Powder Tech. **288**, 229–353 (2016)

5. Barton, N.A.T.: Erosion in Elbow in Hydrocarbon Production System, 1st Edition Crown, Glasgow. (2003)
6. McLaury, B.S., Wang, J., Shirazi, S.A., Shadley, J.R., Rybicki, E.F.: Solid particle erosion in long radius elbows and straight pipes. In: Texas: SPE Annual Technical Conference and Exhibition Proceedings (1997)
7. LNCS Homepage. http://www.diva-portal.org/smash/gets/lncs. Last accessed 08 July 2018
8. LNCS Homepage. http://www.gmigroup.com/products/lncs. Last accessed 28 July 2018
9. LNCS Homepage. http://www.meccanicapadana.it/lncs. Last accessed 25 July 2018